D0603686

The Age of Everything

The Age of Everything

HOW SCIENCE EXPLORES THE PAST

Matthew Hedman

The University of Chicago Press CHICAGO & LONDON

MATTHEW HEDMAN is research associate in the
Department of Astronomy at Cornell University.

The University of Chicago Press, Chicago 60637
The University of Chicago Press, Ltd., London

Printed in the United States of America

16 15 14 13 12 11 10 09 08 07 1 2 3 4 5

ISBN-13: 978-0-226-32292-6 (cloth)
ISBN-10: 0-226-32292-0 (cloth)

Library of Congress Cataloging-in-Publication Data

Hedman, Matthew, 1974–
The age of everything : how science explores the past /
Matthew Hedman.
p. cm.
Includes index.
ISBN-13: 978-0-226-32292-6 (cloth : alk. paper)
ISBN-10: 0-226-32292-0 (cloth : alk. paper) 1. Archaeological dating.
2. Archaeology—Technological innovations. 3. Radiocarbon dating.
4. Earth—Age. 5. Solar system—Age. 6. Science—History.
7. Geochronometry. I. Title.
CC78.H44 2007
930.1—dc22
2006100531

♾ The paper used in this publication meets the minimum requirements of the American National Standard for Information Sciences—Permanence of Paper for Printed Library Materials,
ANSI Z39.48-1992.

CONTENTS

To my Mom and Dad,
and to Judy Burns,
for helping me see this project through.

ACKNOWLEDGMENTS

Many people have helped me with this book over the years. Bruce Winstein and James Pilcher encouraged me to do the series of public lectures that formed the basis on this text. The people in the Enrico Fermi Institute and Kalvi Institute for Cosmological Physics, especially Nanci Carrothers, Charlene Neal, and Dennis Gordon, helped with the practical matters of those lectures. I must also thank all the members of the CAPMAP collaboration, especially Dorothea Samtleben, who put up with me while I worked on the talks and even came to a few lectures. At the end of the lectures, both Bruce Winstein and Christie Henry of the University of Chicago Press convinced me to expand my notes into book form, and the people at Cornell University, especially J. A. Burns and P. D. Nicholson, have been very tolerant of my efforts to do so. In preparing this text, I benefited from comments from Christie Henry, Michael Koplow, and several reviewers.

Todd Telander provided many of the excellent illustrations for this final book. My brother, Kevin, also read through the text and provided many useful suggestions for improving the prose. Throughout all this, my parents, Curt and Sally Hedman provided constant encouragement and support. The following people either provided references and other information for this book or helped me to better understand some of the concepts discussed below: John Harris, the members of the Chicago Maya Society, K. E. Spence, John C. Whittaker, Wen-Hsiung Li, J. David Archibald, Robert Clayton, Stephen Simon, Andrey Kravstov, James Truran, David Chernoff, Ira Wasserman, Stephan Meyer, Erin Sheldon, Rick Kline, and Wayne Hu. Of course, any errors in these pages are my responsibility alone.

CHAPTER ONE

Introduction

From our twenty-first-century perspective, events from the past can often seem impossibly remote. With today's complex technology and constantly shifting political and economic networks, it is sometimes hard to imagine what life was like even a hundred years ago, much less comprehend the vast stretches of time preceding the appearance of humans on this planet. However, thanks to recent advances in the fields of history, archaeology, biology, chemistry, geology, physics, and astronomy, in some ways even the far distant past has never been closer to us. The elegantly carved symbols found deep in the rain forests of Central America, uninterpretable for centuries, now reveal the political machinations of Mayan lords. Fresh interdisciplinary studies of the Great Pyramids of Egypt are providing fascinating insights into exactly when and how these incredible structures were built. Meanwhile, the remains of humble trees are illuminating how the surface of the sun has changed over the past ten millennia. Other ancient bits of wood are helping us better understand the lives of the first inhabitants of the New World. Fossil remains, together with tissue samples from modern animals (including people) suggest that anthropologists may be close to solving the long-standing puzzle of when and how our ancestors started walking on two legs. Similar work might also help biologists uncover how a group of small, shrew-like creatures that lived in the shadow of the dinosaurs gave rise to creatures as diverse as cats, rabbits, bats, horses and whales. The origins of the earth and the solar system are being explored in great detail thanks in part to the rocks that fall from the sky, while the history of the universe can be read in the light from distant stars. The cosmic static that appears on our television sets even allows cosmologists to look back to the very beginning of our universe.

To accomplish all this (and much more besides), scholars and scientists have had to develop a variety of clever ways to figure out when things happened. Without this information, the relics from bygone eras—from impressive stone monuments to humble sticks to feeble starlight—can provide only scattered and almost incomprehensible glimpses of the past. However, once these clues can be arranged and organized in time, the picture becomes much clearer. It becomes possible to evaluate the causes, consequences, nature, and importance of ancient events, and what was once merely an array of random facts takes shape and forms a coherent story.

This book explores how researchers in a wide variety of fields determine the age of things. It grew out of a series of lectures I gave in the spring of 2004 while I was a researcher in the Kavli Institute for Cosmological Physics at the University of Chicago. The talks were part of the Compton Lecture program, which is dedicated to providing the public with information about recent discoveries in the physical sciences. Since at the time I was working as a radio astronomer and cosmologist, it would have been natural for me to discuss the many exciting advances that had taken place in those fields. However, several experts had already given very good lectures on these subjects, and I was encouraged to pursue a different path. I have always been passionately interested in a broad range of academic disciplines—including ancient history, archaeology, evolutionary biology, paleontology, and planetary science—and this gave me an opportunity to offer a multidisciplinary series of talks, each one focusing on a different method of dating ancient objects and events, and how it was being used to revise and reshape our view of the past.

Like the original lectures, this book is not intended to provide an exhaustive catalog of every single dating technique. Nor does it present some sort of comprehensive survey of the history of humanity, the earth, and the universe. Instead, it will focus on a few specific points in time and a sample of methods of measuring age. I hope this approach will allow the reader to gain a deeper understanding of the techniques used in many different fields and to appreciate the special challenges involved in doing research on subjects ranging from the origin of the universe to the politics of the Maya lowlands. In addition, the topics included in this book are all very active areas of study. The following chapters should therefore also provide both background and insight into some of the interesting historical, archaeological, biological, and astronomical discoveries being made today.

However, because the topics covered in this book are still the subjects of active research, it is quite likely that additional discoveries will have come to light by the time you are reading this. Furthermore, several of the topics considered here—such as the colonization of the New World and the use of genetic data to measure time—are still very contentious at the moment. For this reason, I have included lists of articles, books, and websites at the end of each chapter. These should enable curious readers to seek out additional information and perspectives on the issues of interest to them.

I also encourage any interested reader to delve into these references for another reason: I am by no means an expert in all of the subjects covered here. I am well enough versed in topics like ancient history and cosmochemistry to follow the published literature and appreciate technical lectures, but my training is primarily in radio astronomy and observational cosmology. Although I have undergraduate degrees in both physics and anthropology, and even though my current job involves processing data from the Cassini spacecraft in orbit around Saturn, I do not have extensive professional experience in ancient history, archaeology, evolutionary biology, planetary science, and optical astronomy.

I am also well aware of the trouble that can occur when a scientist—particularly a physicist—begins to write about subjects outside their field of expertise. Too often, that scientist seems to be under the mistaken impression that their own training gives them a privileged perspective on a topic others have been studying for decades. I don't want to fall into that trap here, as I have the deepest respect for those who have spent their careers working on these areas. I will therefore tread carefully on other researchers' territories, and point interested readers to those sources that will allow them to further explore any of the subjects covered in this book.

This book begins with events in human history and from there we will move further back in the past all the way back to the Big Bang. Along the way we will cover a broad range of timescales, from mere centuries to billions of years. To help make sense of all this, I have provided the series of time lines in Figure 1.1 to serve as an overview of the events we will consider here.

The time line at the far left of the figure represents the last 100 years, a time span that most of us can readily comprehend and interpret. Marked on the time line are significant events like the two world wars and the moon landings. World War II and the Apollo missions, along with countless moments between and since, are still in people's living memories, but times before this are slowly becoming the domain of history.

Each of the subsequent time lines incorporates fifty times as many years as the time line before it. The second time line thus represents 5,000 years, which includes most of recorded human history. The twentieth century occupies only a tiny fraction of this time. Even the signing of the Declaration of

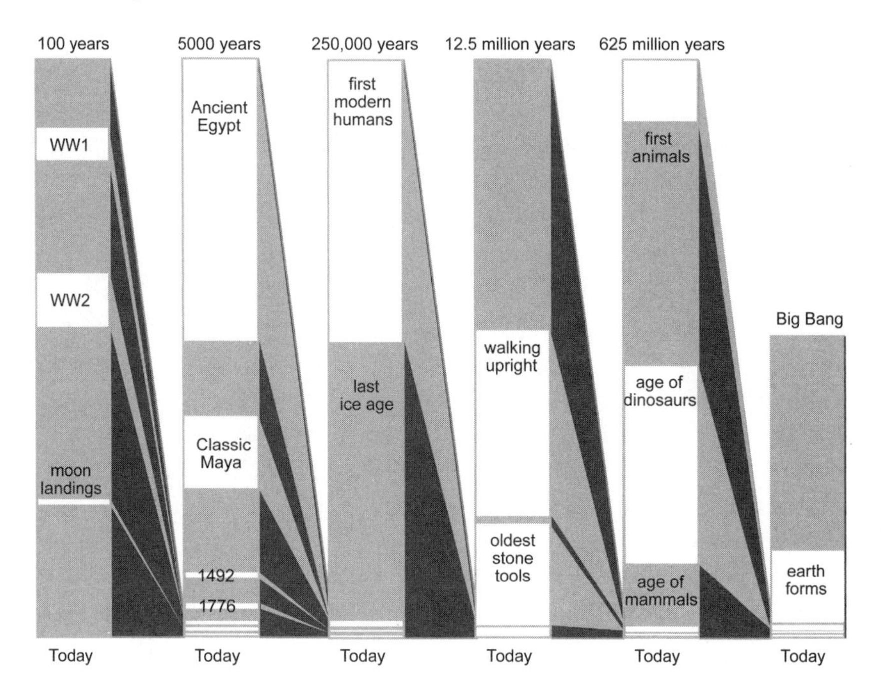

FIGURE 1.1 The timescales of the universe.

Independence in 1776 and Columbus' expedition in 1492 are comparatively recent occurrences on this timescale. Here we come to the first two topics covered in this book: the politics of the Classic Maya civilization of Central America (about 1,500 years ago); and the construction of the Great Pyramids of ancient Egypt (4,500 years ago). Historical records play a crucial role in our understanding of both of these subjects.

Prior to about 5,000 years ago, however, there were no historical records. Scientists therefore must find other means to measure ages. This prehistoric era is covered in the next time line, which represents 250,000 years. Anatomically modern humans—creatures physically indistinguishable from people living today—first appeared about 200,000 years ago, near the top of this time span. This era includes the last great Ice Age and the dispersal of human beings from their earliest home in Africa throughout the rest of the world. For the bottom part of this time span, carbon-14 dating is a key method of measuring ages. A series of three chapters describe this famous dating technique and how it is being used to study such far-flung topics as the physics of the sun and the arrival of people in the New World.

Well before modern *Homo sapiens* made their appearance, there were creatures we would recognize as human-like: they walked on two legs like we do and some even fashioned stone tools. The origins of these traits are included in next time line, which covers a span of 12.5 million years. A combination of fossil evidence and DNA data indicate our ancestors first began to walk upright about 6 million years ago. This pivotal time in our evolution is the subject of chapter 7.

The next time line stretches over 625 million years, encompassing the entire age of dinosaurs and even the origin of multicellular animal life. During this lengthy period, many species arose and went extinct, and the characteristics of life on earth changed in a variety of ways. For example, around the end of the age of dinosaurs, a group of shrew-like animals became the diverse array of creatures we now call mammals. As we will see in chapter 8, analyses of the DNA of living animals may be able to shed new light on this remarkable transformation.

Note that the final time line is shorter than the rest. Were it extended to the same length as the others, it would represent 31.25 billion years. Our universe is not that old, so this bar has been shrunk down to begin at the Big Bang, which occurred less than fifteen billion years ago. Well after this point on the time line, we can see the formation of the earth and the solar system, which is the subject of chapter 9. Voyaging even deeper into the past, the last three chapters will discuss the age of the oldest known stars and even the birth of the universe itself.

In addition to this visual depiction of the history of the universe, here are some useful rules of thumb to help keep these various timescales straight:

- Recorded history is about twenty times as long as the history of the United States.
- Humans have been around about forty times as long as recorded history.
- The ancestors of humans have been walking upright about thirty times as long as modern humans have been around.
- The last giant dinosaurs are about ten times as old as the first ancestor of humans that walked upright.
- The first multicellular animals are about ten times as old as the last giant dinosaurs.
- The earth and solar system are about eight times as old as the first multicellular animals.
- The universe is about three times as old as the earth and the solar system.

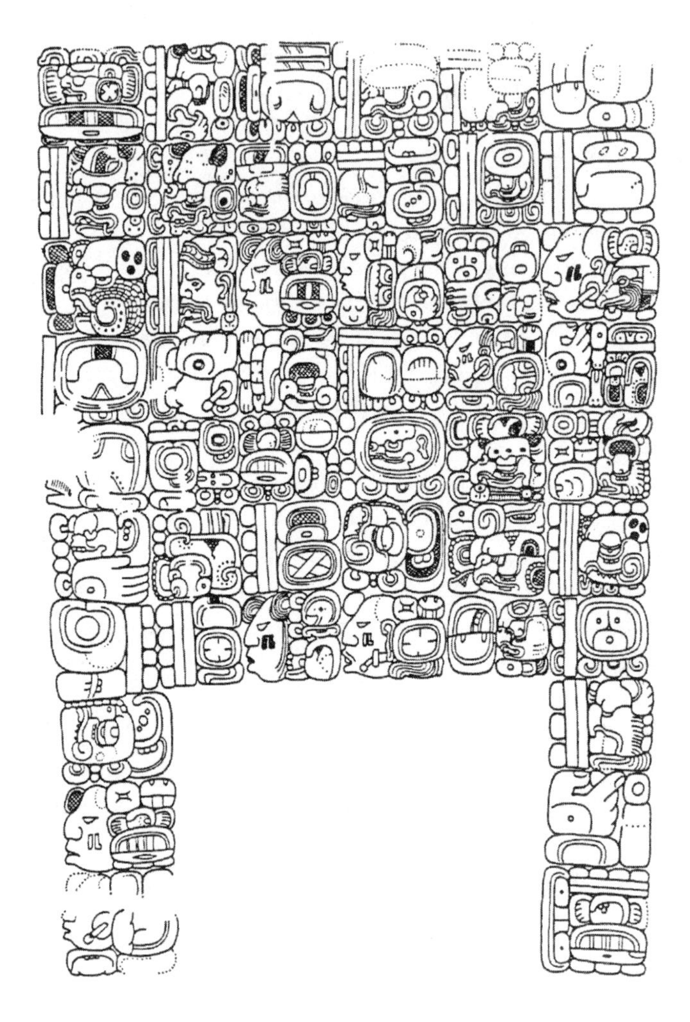

FIGURE 2.1 A Mayan text, from Piedras Negras Stela 3 (drawn by L. Schele © David Schele, courtesy Foundation for the Advancement of Mesoamerican Studies Inc., www.famsi.org).

CHAPTER TWO

The Calendars of the Classic Maya

We begin this journey through the past in the lowlands of the Yucatan Peninsula, a region of Central America now split between Guatemala, Belize, western Nicaragua, and southeastern Mexico. Here, scattered throughout the tropical forest, are hundreds of stones and buildings inscribed with various combinations of human body parts, bits of animals, pieces of plants, and a host of more abstract symbols (see Figure 2.1). For hundreds of years, these inscriptions were either almost unknown to the outside world or virtually incomprehensible. However, during the last century it became clear that these carvings are in fact a unique form of writing, and over the last thirty years scholars have managed to decipher a large number of the hieroglyphic signs in these texts. This research has revealed that most of the inscriptions were written over one thousand years ago—during what is now known as the Classic period (250–900 CE)—by a people who spoke a Mayan language. Indeed, these monuments were probably made by some of the ancestors of the Mayan people who still live in the area today.

Each year, new texts are discovered, shedding more light on the religion, culture, and politics of these ancient Mayans. In 2001, archaeologists working in a ruined city now called Dos Pilas found parts of a stone staircase covered with hieroglyphs. The text on one of the steps read, in part:

> war came upon [a place associated with Dos Pilas], supervised by Yuknoom Ch'een, the holy lord of Calakmul. B'alah Chan K'awiil, the holy lord of Dos Pilas, escaped, going down to a place called K'inich Pa' Witz.

FIGURE 2.2 An example of a Classic Mayan date. From Piedras Negras Stela 3 (drawn by Linda Schele).

The meaning of this inscription is clear: Dos Pilas was attacked by another city, and its king was forced to flee. However, the significance of this conflict is not so easy to grasp. Who were Yuknoom Ch'een and B'alah Chan K'awiil? What led them to fight each other? What were the consequences of this battle? This short inscription alone cannot answer these questions. There are other texts that record the deeds of both of these kings that could help reveal the importance of this battle, but we still need to figure out how this one conflict relates to the events and activities recorded elsewhere.

Fortunately, like most Classic Mayan texts, the above inscription is associated with a series of hieroglyphic signs that provide an exact date using one of the most sophisticated and elaborate calendars ever devised. Such a precise date can tell us how this battle fits into the lives of Yuknoom Ch'een and B'alah Chan K'awiil, and similar dates have allowed modern Mayanists to sketch out the biographies of Yuknoom Ch'een, B'alah Chan K'awiil, and dozens of other Mayan kings and nobles who lived and died centuries before Columbus reached the New World.

SECTION 2.1: THE MAYAN CALENDAR

The intricacies of the Mayan calendar are best illustrated using an example like the date shown in Figure 2.2. Like all Mayan writing, this date is written in a series of square blocks composed of multiple hieroglyphs. Within each block, the signs are typically read starting at the upper left and progressing through to the lower right, although there are plenty of exceptions to this pattern. The blocks themselves are read in a rather strange order. The texts are organized in pairs of columns, and the reader starts at the upper left of each column pair, reads across both columns, then moves down to the next row until reaching the bottom of the text and continuing to the next pair of columns.

As	An example	In	The same
This	Sentence	Sort of	Pattern
Is	Written	As	We find
With	The words	In Mayan	texts

Following this conventional reading order, this date starts with the block at the upper left:

FIGURE 2.3

Although the symbol in the center of this block does contain some calendrical information, its primary function is to signal that there is a date in the following hieroglyphs.

The five signs following the initial sign form something called the Long Count. For clarity, we can lay out these signs in a familiar left-to-right order:

FIGURE 2.4

Each of these signs consists of some combination of bars and oval-shaped dots in front of the head of a very odd-looking creature. In spite of the elaborate details carved into the heads, it is the bars and dots that contain most of the information. Each combination of bars and dots represents a number between 0 and 19. Each dot corresponds to a 1, while each bar corresponds to a 5. At the left edge of the first glyph are a column of four oval dots and a vertical bar; together these make 9. In the second glyph there are two bars and two dots separated by a (numerically meaningless) spacer shaped like a sideways U, and so it has a value 12. The third glyph has only two dots (between two spacers) and therefore corresponds to our 2. The fourth glyph does not have any bars or dots. In their place there is a sign consisting of a disc surrounded by three petals, a symbol known to be equivalent to our zero. The final sign, with three bars and one dot framed by two spacers, has a value of 16. Conventionally, Mayanists write out this sequence of numbers like this: 9.12.2.0.16.

The heads attached to these numbers indicate how they should be combined to express a single count of days. However, for our purposes, this information is redundant because the order of the numbers also establishes the total number of days. In this respect, the Long Count has some remarkable similarities with our own counting system. Remember that our number 482 represents a combination of 4 hundreds, 8 tens, and 2 ones. The Long Count uses a similar system of "places" to express large numbers, but there is one important difference between modern numbers and Mayan Long Counts. In our numbers each numeral corresponds to a different power of 10: 1, 10, 100, 1,000, and so on. By contrast, in the Mayan Long Count the numbers correspond to units of 1, 20, 360, 7,200 and 144,000. The logic of this sequence becomes more obvious if it is rewritten as a series of products: 1, 20, 18 × 20, 20 × 18 × 20, and 20 × 20 × 18 × 20. Thus, the Long Count mainly uses powers of 20 instead of powers of 10, which is a perfectly reasonable alternative to our own system. The 18 introduced in the third term is clearly the exception to this pattern, and this occurs because the Long Count is a number of days, and 360 = 18 × 20 is the closest multiple of twenty to the 365.25 days in a year. Using this system, we can calculate that the long count 9.12.2.0.16 corresponds to the following number of days (note that, like in our own system, the first number corresponds to the largest unit):

$$(9 \times 144{,}000) + (12 \times 7{,}200) + (2 \times 360) + (0 \times 20) + 16 = 1{,}383{,}136 \text{ days}$$

This is nearly 3,800 years, so "Long Count" is not a misnomer, and it stands to reason that this number indicates how many days have passed since some important ancient event. Fortunately, elements of this system were still being used when the Spanish made contact with the Mayans, and documents from this much later time period allow scholars to coordinate the Mayan day count with our own calendar. These analyses suggest that the Long Count date 0.0.0.0.1 corresponds to sometime in August during the year 3114 BCE. This is millennia earlier than any known text, so any historical interpretation of this date is suspect. However, such ancient times are mentioned on numerous occasions in mythological texts, so this date did have a deep religious significance to the Maya. At a more practical level, since Long Counts always uses this same ancient date as a reference point, they specify exactly when any event happened. For example, the five numbers given above refer to July 7, 674 CE.

In principle, the Long Count provides all the information either the Mayans or modern Mayanists would need to determine exactly when any event happened. However, the Long Count is only one part of a complete Mayan date. The date in Figure 2.2, for example, contains eight blocks in addition to the Long Count. Most of these hieroglyphs provide information about the phase of the moon and the status of the lunar month, elements of the Mayan calendar that will not be discussed in detail here. Instead, let us focus on the sign immediately following the Long Count and the last sign in the date:

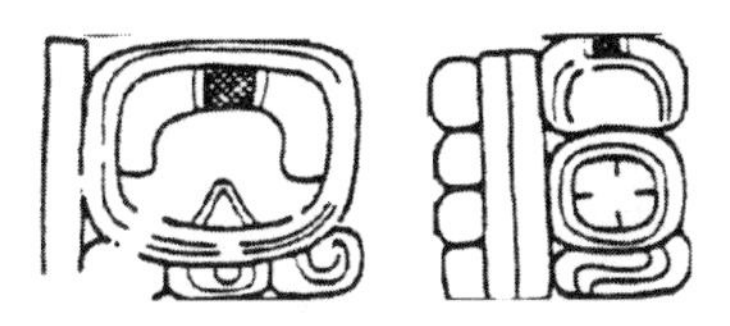

FIGURE 2.5

FIGURE 2.6 The twenty possible day signs that can occur in the Tzolk'in. (The order in this figure is vertical; e.g., Ik' follows Imix, Chikchan follows K'an, and so on.)

Individually, these signs are known as the *Tzolk'in* and the *Haab*, respectively. Together, they are called the Calendar Round and they are often used as a shorthand for a complete date.

The Tzolk'in has two parts: a number and a day sign. As in the Long Count, the number is represented by bars and dots, but this number can have a value only from 1 to 13. In this case, it is 5. There are twenty possible day signs in the Tzolk'in, each of which is represented by a unique figure carved into the oval (see Figure 2.6). In this example, we have the sign known as Kib. This Tzolk'in date is therefore 5 Kib. Both the number and the day sign of the Tzolk'in change every day. With each new day, the number increases by one until it reaches 13, at which point it cycles back to one. At the same time, the day sign changes through a particular sequence of its twenty possible values. Thus the next Tzolk'in date after 5 Kib is 6 Kaban, followed by 7 Etz'nab, 8 Kawak, 9 Ahaw, 10 Imix, 11 Ik', and so on. Note that 13 and 20 are not multiples of each other, so the next time the number is 5 the day sign will not be Kib but rather Muluk. The same combination of number and day sign occurs only every 260 days.

The Haab, on the other hand, is a cycle of 365 days made up of eighteen "months" of twenty days (numbered from 0 to 19) each plus one short "month" with five days (Figure 2.7). The number here indicates the day of the month and the sign attached to it is the name of the month. In this example we have the number 14 attached to the sign for the month Yaxk'in. The following day would then be 15 Yaxk'in, then 16 Yaxk'in, and so on until we reach 19 Yaxk'in, after which we have the day 0 Mol.

Both the Tzolk'in and the Haab advance every day, so all components of the Calendar Round are constantly changing. One popular way to visualize the working of this date uses gears, as shown in Figure 2.8. Each day the gears all shift over one unit to produce the Calendar Round date for that day.

Both the Tzolk'in and Haab are rather short cycles and specify when something happened only to within a year. However, since the Tzolk'in is a 260-day cycle and the Haab is a 365-day cycle, the same combination of Tzolk'in and Haab occurs only once every 18,980 days (260 × 365 = 94,900, with the common factor of 5 divided out), which is roughly fifty-two years.

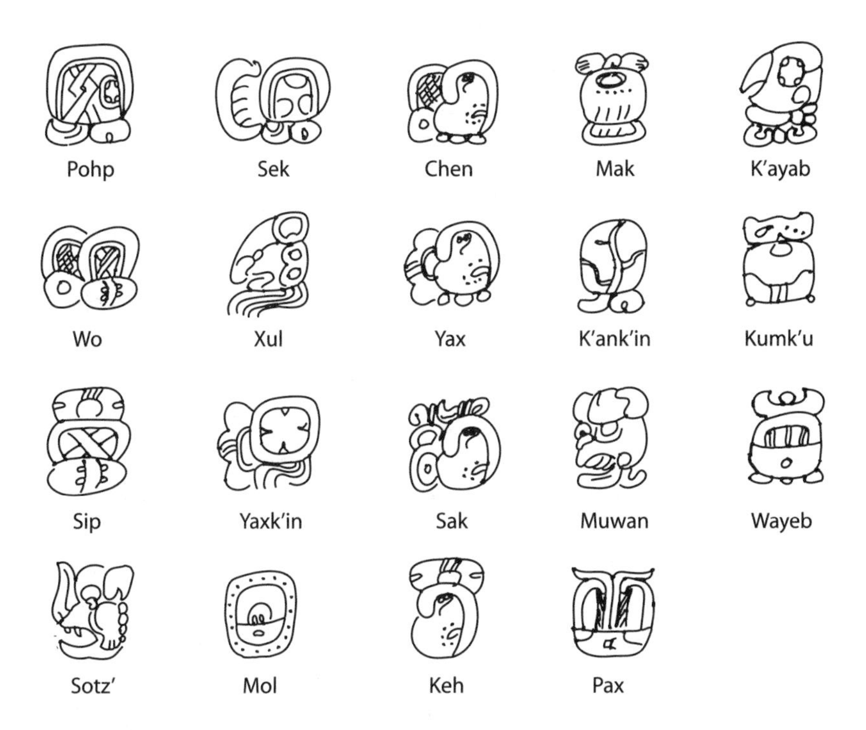

FIGURE 2.7 The month signs of the Haab. (The order in this figure is vertical; e.g., Wo follows Pohp, Sek follows Sotz', and so on.)

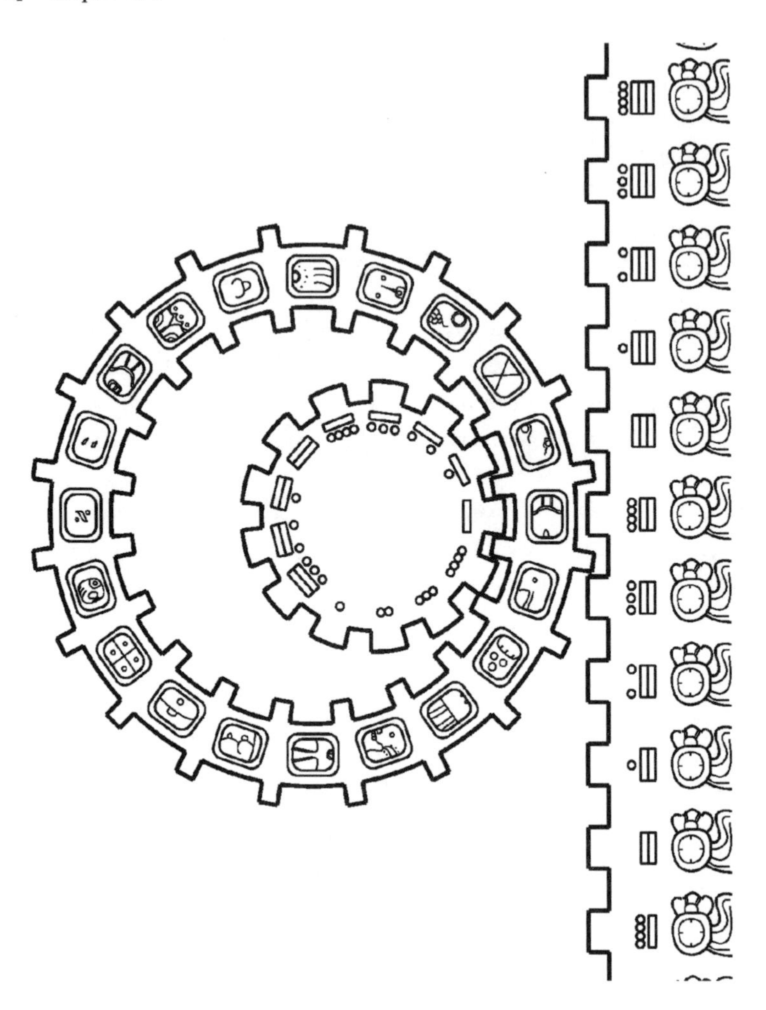

FIGURE 2.8 The mechanics of the Mayan Calendar Round illustrated using gears, one each for the number and day sign of the Tzolk'in and one for the Haab. The current day corresponds to where the gears are aligned. Every new day the gears advance together by one position.

This means that if we know the Calendar Round for an event, we know exactly on what day it happened within a given fifty-two-year cycle, so it is possible to extract the exact date for an event from the Calendar Round with only a partial Long Count or a reference to a king who lived during a particular cycle. Conversely, the Calendar Round date can be reconstructed if only the Long Count is known. Therefore, even if the inscription is partially eroded, the remaining parts of the Long Count and the Calendar Round often contain enough chronological information to determine when the recorded events occurred.

SECTION 2.2: THE MAYAN CALENDAR AND THE NATURE OF THE MAYAN TEXTS

The elaborate Mayan calendar has been a powerful tool for modern Mayanists because almost every single event the Mayans recorded on their stone monuments is dated with either a Calendar Round or a Long Count. We can therefore compute, almost to the day, when each and every recorded event happened. This chronological information has proven invaluable to many efforts to understand the life and history of the Classic Maya. In fact, these dates even played a pivotal role in establishing the true nature and content of the Classic Mayan inscriptions. In the 1960s, Tatiana Proskouriakoff studied the texts from the city of Piedras Negras and found six distinct groups of texts, each associated with a particular location in the site. Each of these groups contained dates spanning several decades, and the first two events recorded in a given group were almost always indicated by the following signs:

FIGURE 2.9

The first sign is sometimes called the "up-ended frog" (turn the page sideways to see the frog's head), while the second sign is known as the "toothache" glyph because the knotted element looks something like a bandaged tooth. In the groups studied by Proskouriakoff, the "up-ended frog" event always occurred a few decades before the "toothache" event. All of the other events recorded in a given group happened after the "toothache" event, and the final event in the group always occurred before the next "toothache" event is recorded in one of the other groups.

Proskouriakoff studied these suspicious temporal relationships and reasoned that the patterns she found could be explained if the events in each group, which cover a time period comparable to a human life span, refer to events in the life of a particular ruler of Piedras Negras. This means the "up-ended frog," as the first event in any given sequence, marks the birth of the ruler, and the "toothache" event indicates the ruler's coronation when he was in his twenties or thirties. The time span covered by the other events then corresponds to the rule of the king.

Further study of the last events associated with various rulers turned up another pair of signs:

FIGURE 2.10

When both of these symbols are present, the event indicated by the skull is always noted as taking place a matter of days or weeks before the other event. Since these events occurred close to the coronation of the next ruler, they almost certainly mark the death of the current ruler and his burial, respectively. The use of a skull to mark the former event is consistent with such an interpretation.

Proskouriakoff's analysis clearly demonstrated that the Mayan texts contain historical information. This insight spurred a renewed interest in the Mayan texts as historical documents. Furthermore, by ascribing specific meanings to certain hieroglyphs, this work aided in efforts to translate the Mayan inscriptions. The Mayan calendar therefore has played a crucial part in Mayan historical studies since the very beginning.

SECTION 2.3: THE LIFE AND TIMES OF YUKNOOM CH'EEN

Just as the Mayan calendar helped illuminate the true nature of the Mayan inscriptions in the 1960s, today it is being used to unravel the complex political lives of the Classic Mayan elite. The Mayans of the Classic period do not appear to have ever been unified under a single political entity, but instead lived in a multitude of more or less independent cities, some of which are shown in Figure 2.11. These cities and their rulers interacted through a tangled web of rituals, diplomacy, marriage, and warfare. This bewildering jumble of events can be sorted out only because a date is recorded for almost

FIGURE 2.11 The region of Central America occupied by the Classic Maya, with some of the more important cities indicated by their modern names.

every single battle and ritual. These events can therefore be arranged in chronological order, allowing Mayanists to evaluate and explore their causes and consequences in great detail.

One influential Mayan whose life story has recently begun to emerge from the texts is Yuknoom Ch'een, one of the kings mentioned in the inscription from the beginning of this chapter. This lord (whose portrait and name are shown in Figure 2.12) ruled a city now known as Calakmul—it was called Chan, or "Snake," by the Classic Maya—but he had an impact on cities throughout the Mayan area, and his activities are recorded on texts scattered throughout the Yucatan. At the same time, the tropical forest has done considerable damage to many of Calakmul's monuments, so most of the texts Yuknoom Ch'een himself commissioned to chronicle his life are too eroded to read. The few texts surviving in and around Calakmul tell us only that Yuknoom Ch'een was born in 600 CE; that he became ruler of Calakmul at the age of thirty-six, and that he probably died circa 686. The records from other Mayan cities provide most of the available information about Yuknoom Ch'een's activities, and the dates associated with these events are especially important in putting together a coherent story of his life.

FIGURE 2.12 Yuknoom Ch'een's name (left) and portrait (right). Redrawn from Simon Martin and Nikolai Grube *Chronicle of the Maya Kings and Queens* (Thames and Hudson, 2000), 108.

Figure 2.13 and the appendix to this chapter summarize the known events of Yuknoom Ch'een's long life. Yuknoom Ch'een was born during the rule of another important king of Calakmul, who is called Scroll Serpent today because his real name has not yet been translated. When Yuknoom Ch'een was eleven years old, Scroll Serpent launched a memorable attack on Palenque, located nearly three hundred kilometers to the west. Sacking a city so far away was an impressive feat, and this was actually the second time Calakmul had struck Palenque, a previous attack having occurred in 599. This second campaign had a great impact on Palenque and was recalled in inscriptions written over seventy years after the fact. In all likelihood, this event was also celebrated at Calakmul as an example of the great power wielded by the city and its king.

Scroll Serpent died sometime during the next few years, roughly twenty years before Yuknoom Ch'een ascended to the throne. Events during this intervening time were documented at a site called Caracol, which is located some two hundred kilometers southeast of Calakmul. Three lords of Calakmul are mentioned in the Caracol records from this time: Yuknoom Chan, Tajoom Uk'ab K'ak', and Yuknoom Head. Each of these kings ruled for less than ten years. While such a series of short reigns is not particularly unusual, Yuknoom Head was also involved in some sort of conflict just two months before his reign ended and Yuknoom Ch'een assumed the throne. This coincidence, coupled with the fact that Yuknoom Ch'een was over thirty-five years old when he became king, suggests that there was some turbulence in Calakmul's royal house at this time.

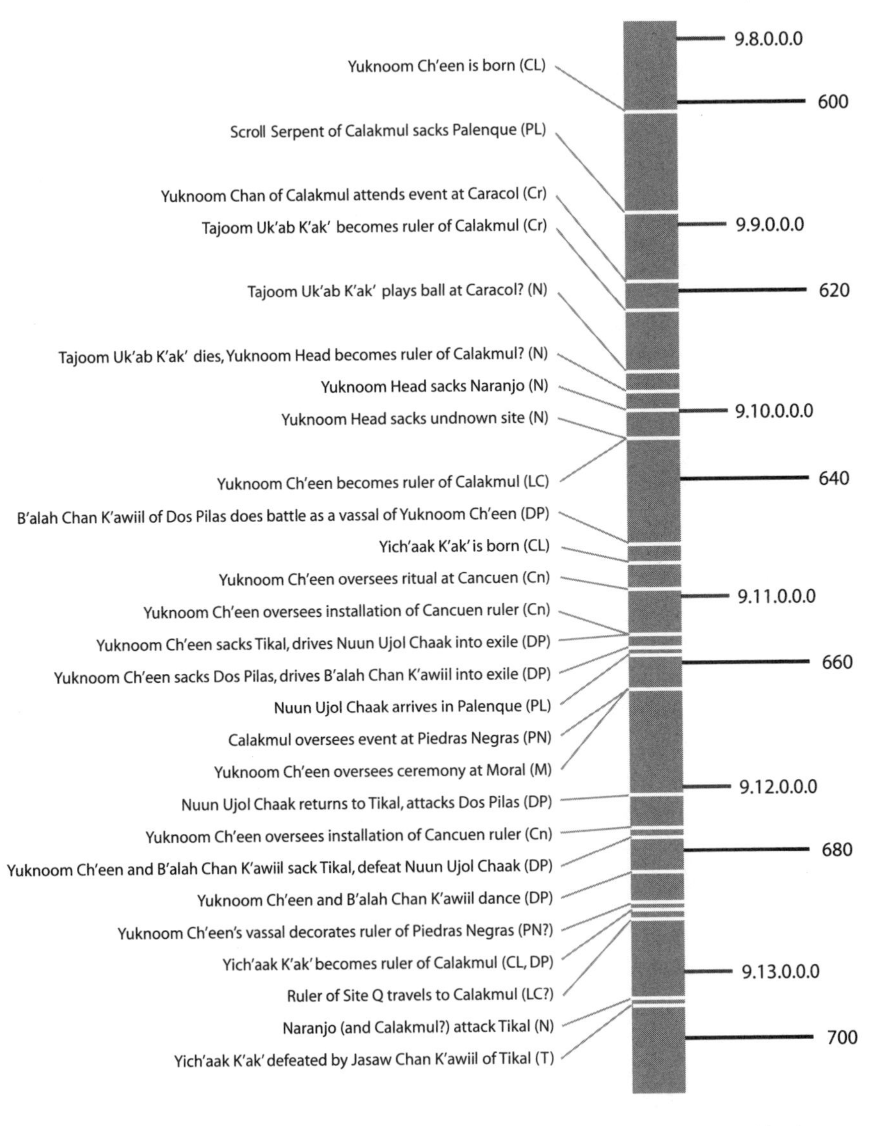

FIGURE 2.13 A summary of the significant events in the life of Yuknoom Ch'een.

If Calakmul was having problems when Yuknoom Ch'een began his reign, they must have been resolved by the end of his first decade on the throne. About ten years after he became ruler, Yuknoom Ch'een is mentioned at cities well to the south of Calakmul such as Cancuen and Dos Pilas. Yuknoom Ch'een's activities in this region demonstrate not only that Calakmul could still operate in distant cities, but also that Calakmul was able to influence politics in the south without much interference from its greatest rival, Tikal.

Tikal—known to the Classic period Mayans as Mutul—was a city whose power over other Mayan sites often rivaled Calakmul's. Indeed, much of Mayan politics revolved around the changing fortunes of these two centers. During Yuknoom Ch'een's reign the ruling house of Tikal appears to have fallen on hard times, for two people claimed to be "ruler of Tikal": Nuun Ujol Chaak, who ruled from Tikal itself, and B'alah Chan K'awiil, who founded a "new Tikal" at the site now called Dos Pilas. Yuknoom Ch'een seems to have taken full advantage of the schism between these two lords, using it as an opportunity to extend Calakmul's influence to the south. Indeed, Yuknoom Ch'een (or perhaps his representatives) was able to travel several times over a period of twenty-five years to the southern city of Cancuen to oversee rituals and the coronation of rulers. On each of these trips, Yuknoom Ch'een was able to pass through Tikal's territory without any reported harassment.

Yuknoom Ch'een did more than just watch the feud develop between the lords of Tikal; he actively engaged and exacerbated the conflict. Until recently, the available texts indicated that Yuknoom Ch'een quickly forged a lasting alliance with B'alah Chan K'awiil against Nuun Ujol Chaak. During Yuknoom Ch'een's twelfth year on the throne, B'alah Chan K'awiil—who was described as a vassal of the king of Calakmul—was involved in some sort of conflict with an obscure person from Tikal. In later texts, Yuknoom Ch'een and B'alah Chan K'awiil worked together several times against Nuun Ujol Chaak, the "true" lord of Tikal.

This picture of a stable, natural alliance between the kings of Calakmul and Dos Pilas was shattered by the discovery of the inscription quoted at the beginning of this chapter, which says that Yuknoom Ch'een attacked Dos Pilas and drove B'alah Chan K'awiil into exile. This happened in Yuknoom Ch'een's twenty-second year as ruler, just two years after a campaign in which he sacked Tikal and drove Nuun Ujol Chaak into exile. During this time, Calakmul was apparently hostile to both the lords of Tikal. The experts are still not entirely sure what to make of this new information. Perhaps B'alah Chan K'awiil and Yuknoom Ch'een had a falling out, or maybe B'alah Chan K'awiil's earlier connection with Calakmul was just a fiction introduced retroactively into the texts. We can only hope that new texts and future research will eventually clarify the relationship between Calakmul and Dos Pilas during these early years.

While the prior allegiances of the lords of Tikal is uncertain, after their defeat at the hands of Yuknoom Ch'een the political situation becomes clearer. Sometime after they were driven out of their respective cities, both Nuun Ujol Chaak and B'alah Chan K'awiil were involved in a ritual with Yuknoom Ch'een's successor. While the relevant text is unfortunately so heavily eroded that the exact nature of this event is uncertain, it likely involved both the lords of Tikal swearing loyalty to Calakmul. B'alah Chan K'awiil apparently accepted this, and remained an ally of Calakmul from that point on. Nuun Ujol Chaak, however, seems to have had other ideas.

About two years after he was forced out of his city, Nuun Ujol Chaak arrived at Palenque, the city that Scroll Serpent had sacked nearly fifty years earlier. At about the same time, cities in the same region, such as Moral and Piedras Negras, record that Yuknoom Ch'een and his representatives were present to supervise various rituals. This is a rather suspicious coincidence, and it is easy to imagine that the rivalry between these two kings was guiding their interests in this region. Nuun Ujol Chaak could have been rallying support for his cause from Palenque, a city that had also suffered at the hands of Calakmul, and also happened to be situated far enough away to serve as a secure base of operations. Yuknoom Ch'een, on the other hand, may have been forging alliances with Moral and Piedras Negras in order to isolate Palenque from other groups sympathetic to Tikal.

Geopolitical maneuvering, if that is what it was, eventually gave way to open conflict when, fifteen years after his exile began, Nuun Ujol Chaak finally returned to the Tikal area and managed to drive B'alah Chan K'awiil out of Dos Pilas. However, Nuun Ujol Chaak's victory was short lived, since five years later B'alah Chan K'awiil and Yuknoom Ch'een together expelled Nuun Ujol Chaak from Tikal; a little later, B'alah Chan K'awiil defeated and likely killed him. At this time Yuknoom Ch'een would have been over seventy-seven years old and surely did not take part in any battle directly.

Yuknoom Ch'een's old age does not seem to have compromised Calakmul's widespread influence. He and his representatives continued to participate in various ceremonies in far away cities like Dos Pilas and Piedras Negras even as he neared his fiftieth year on the throne. This great king of Calakmul probably died in 686 as an octogenarian, for that is when his successor rose to power. The new ruler of Calakmul, a man named Yich'aak K'ak', would not fare as well against the forces of Tikal and its energetic new lord, Jasaw

Chan K'awiil, but these events belong to a different chapter in the history of the Maya.

After this survey of the events in Yuknoom Ch'een's life, we can now begin to appreciate how powerful accurate dates can be for reconstructing the history of the Classic Maya. We would not know about Yuknoom Ch'een's late arrival on the throne of Calakmul, the complicated nature of his relationship with Dos Pilas, or the possible motives behind his later political activities in the west if we did not have the precise temporal information encoded in the Mayan texts. The Mayan calendar therefore greatly improves our understanding of Yuknoom Ch'een's life. Indeed, it enriches all of Classic Mayan history. As we will see in the next chapter, many people were not as fastidious about recording the dates of events as the Mayans. Thus historians usually have to work much harder to arrange and organize events in time, using a variety of more indirect, and sometimes surprising, indicators of age.

SECTION 2.4: FURTHER READING

Two of the most useful websites dealing with the Classic Maya and Mesoamerica are www.famsi.org and www.mesoweb.com. The former contains a report by Frederico Fahsen (www.famsi.org/reports/01098/index.html) describing the discovery of the text translated at the beginning of this chapter.

The Mayan calendar is dealt with in most books on the Mayan script, such as: Michael D. Coe and John D. Stone, *Reading the Maya Glyphs* (Thames and Hudson, 2001), John F. Harris and Stephen K. Stearns, *Understanding Mayan Inscriptions,* 2nd ed. (University of Pennsylvania Museum Press, 1997), and John Montgomery, *How to Read Mayan Hieroglyphs* (Hippocrene Books, 2002).

For a history of Mayan decipherment, including Proskouriakoff's work, see Michael D. Coe *Breaking the Maya Code* (Thames and Hudson, 1992).

A very thorough and reasonably up-to-date review of Mayan history is Simon Martin and Nikolai Grube *Chronicle of the Maya Kings and Queens* (Thames and Hudson, 2000).

APPENDIX: SOURCES FOR LIFE AND TIMES OF YUKNOOM CH'EEN

For the curious reader, the texts that provide information on the life of Yuknoom Ch'een are as follows.

Long Count	Calendar Round	Gregorian Date	Event	Source
9.8.7.2.17	8 Kaban 5 Yax	11 Sep 600	Yuknoom Ch'een born	Calakmul Stela 33
9.8.17.15.14	4 Ix 7 Wo	4 Apr 611	Scroll Serpent sacks Palenque	Palenque Temple of the Inscriptions
9.9.5.13.8	4 Lamat 6 Pax	6 Jan 619	Yuknoom Chan of Calakmul oversees event at Caracol	Caracol Stela 3
9.9.9.0.5	11 Chikchan 3 Wo	28 Mar 622	Tajoom Uk'ab K'ak' becomes ruler of Calakmul	Caracol Stela 22
9.9.15.3.10	13 Ok 18 Sip	30 Apr 628	Tajoom Uk'ab K'ak' performs ritual (at Caracol?)	Naranjo Stairway
9.9.17.11.14	13 Ix 12 Sak	1 Oct 630	Tajoom Uk'ab K'ak' dies	Naranjo Stairway
9.9.19.16.3	7 Ak'bal 16 Muwan	24 Dec 631	Yuknoom Head of Calakmul attacks and defeats Naranjo	Naranjo Stairway
9.10.3.2.12	2 Eb 0 Pohp	4 Mar 636	Yuknoom Head attacks and defeats unknown site	Naranjo Stairway
9.10.3.5.10	8 Ok 18 Sip	28 Apr 636	Yuknoom Ch'een becomes ruler of Calakmul	La Corona Altar
9.10.15.4.9	4 Muluk 2 Kumk'u	4 Feb 648	B'alah Chan K'awiil of Dos Pilas acts as vassal of Yuknoom Ch'een	Dos Pilas Stairway 4
9.10.19.5.14	3 Ix 7 Kumk'u	8 Feb 652	Yuknoom Ch'een oversees event at Cancuen	Cancuen Panel
9.11.4.4.0	11 Ahaw 8 Muwan	9 Dec 656	Yuknoom Ch'een oversees installation of Cancuen ruler	Cancuen Panel
9.11.4.5.14	6 Ix 2 K'ayab	12 Jan 657	Yuknoom Ch'een attacks Tikal, and drives its ruler, Nuun Ujol Chaak, into exile	Dos Pilas Stairway 2

Long Count	Calendar Round	Gregorian Date	Event	Source
9.11.6.4.19	9 Kawak 17 Muwan	18 Dec 658	Yuknoom Ch'een attacks Dos Pilas and drives its ruler, B'alah Chan K'awiil, into exile	Dos Pilas Stairway 2
9.11.6.16.17	13 Kaban 10 Chen	16 Aug 659	Nuun Ujol Chaak arrives in Palenque	Palenque Temple of the Inscriptions
9.11.9.8.6	12 Kimi 9 Kumk'u	7 Feb 662	Calakmul oversees event in Piedras Negras	Piedras Negras Stela 35
9.11.9.11.3	4 Ak'bal 1 Sip	5 Apr 662	Yuknoom Ch'een oversees installation of Moral ruler	Moral Stela 4
9.12.0.8.3	4 Ak'bal 11 Muwan	8 Dec 672	Nuun Ujol Chaak returns and attacks Dos Pilas	Dos Pilas Stairway 2
9.12.4.11.1	7 Imix 9 K'ayab	14 Jan 677	Yuknoom Ch'een oversees installation of Cancuen ruler	Cancuen Panel
9.12.5.10.1	9 Imix 4 Pax	20 Dec 677	Yuknoom Ch'een and B'alah Chan K'awiil attack Nuun Ujol Chaak	Dos Pilas Stairway 4
9.12.12.11.2	2 Ik' 10 Muwan	4 Dec 684	Yuknoom Ch'een and B'alah Chan K'awiil perform ceremony together	Dos Pilas Stairway 2
9.12.13.4.3	2 Ak'bal 6 Mol	13 July 685	Representative of Yuknoom Ch'een decorates ruler of Piedras Negras	Piedras Negras panel
9.12.13.17.7	6 Manik 5 Sip	3 Apr 686	Yich'aak K'ak' becomes ruler of Calakmul	Calakmul Stela 9

And here are where these texts can be found:

Calakmul: Line drawings of Stelae 9 and 33 are not easily available, but pictures of the stelae are published by Karl Ruppert and John H. Denison, Jr. in *Archaeological Reconnaissance in Campeche, Quintana Roo, and Peten* (Carnegie Institute of Washington, publication 543, 1943).

Cancuen: "Looted Panel." Line drawing appears at www.mesoweb.com/features/cancuen/index.html, "A Reading of the Cancuen Looted Panel" by Stanley Guenter.

Caracol: Stela 3 is published in Carl P. Beetz and Linton Satterwaite *The Monuments and Inscriptions of Caracol, Belize* (University Museum Publications, 1982). Stela 22 is published in PARI Monograph 7 (Studies in the Archaeology of Caracol, Belize, 1994), in "Epigraphic Research at Caracol, Belize" by Nikolai Grube.

Dos Pilas: Hieroglyphic Stairway 4 is published in Stephen D. Houston *Hieroglyphs and History of Dos Pilas* (University of Texas Press, 1993); Hieroglyphic Stairway 2 is available through www.famsi.org/reports/01098/index.html. Translations of the two texts are available at www.mesoweb.com/features/boot/DPLHS2.html www.mesoweb.com/features/boot/DPLHS4.html in articles by Erik Boot.

La Corona: Line drawings of the altar are not readily available to my knowledge (see also Martin and Grube *Chronicle of the Maya Kings and Queens*).

Moral: Stela 4, published in Simon Martin "Moral: Reforma y la Contienda por el Oriente de Tabasco" in *Arqueologia Mexicana* 9, no. 61 (2003): 44–47.

Naranjo: Hieroglyphic Stairway, published in Ian Graham *Corpus of Mayan Hieroglyphic Inscriptions,* vol. 1 (Peabody Museum, 1975–).

Palenque: A line drawing of the Temple of the Inscriptions is found in Linda Schele and Peter Matthews *The Code of Kings* (Scribner, 1998).

Piedras Negras: Stela 35 published in Nikolai Grube "Palenque in the Maya World" for the Eighth Palenque Round Table, available on www.mesoweb.com/pari/publications/RT10/001grube/text.html.

CHAPTER THREE

Precession, Polaris, and the Age of the Pyramids

Thousands of years before the Mayans carved ornate inscriptions on the walls of their buildings, the Egyptians constructed some of the most famous and impressive monuments of the ancient world: the Great Pyramids. Built of millions of blocks of stone, some weighing over fifty tons, these enormous tombs are feats of ancient engineering. Scholars have been investigating Egypt's pyramids for well over a hundred years now, and this research has yielded much information about both the structures themselves and the people responsible for them. However, the pyramids have not yet given up all of their secrets, and in spite of all this research, we still do not know precisely when these massive structures were built.

Unlike the inscriptions of the Classic Maya, which include an exceptional amount of calendrical information, Egyptian hieroglyphic texts usually document only the number of years that the current king has been on the throne. For example, inscriptions recording the construction of a certain pyramid merely mention a "year of the fifteenth occasion" of a king named Snofru, referring to a series of events which occurred every one or two years during each king's reign. This date therefore only tells us that the work occurred some number of years after Snofru became ruler of Egypt, hardly enough information to establish how long ago the pyramid was constructed.

In the absence of explicit historical records, Egyptologists seeking to determine the age of the pyramids must rely on a combination of historical, archaeological, and even astronomical data. At present, this sort of information indicates that the biggest of the Great Pyramids was built between 4,400 and 4,600 years ago. This sort of age estimate is enough to convey the extreme antiquity of the pyramids, but we must also remember that 200 years is a very long time compared to a human life span. The pyramids today would be equally awe-inspiring if they were built 4,400 or 4,600 years ago, but an Egyptian who was alive while the pyramids were being constructed would certainly have a different view of these structures than someone who was born 200 years after they were finished. A more precise date for the pyramids would therefore allow us to better understand the impact these tremendous construction projects had on Egypt's people and history. Recently, a novel method of combining astronomical and archaeological data has been proposed that may be able to pinpoint the age of all of the Great Pyramids to within a few years.

SECTION 3.1: THE PYRAMIDS AND THE HISTORY OF ANCIENT EGYPT

The first step towards discovering the age of the pyramids is to determine whom they were built for. Each pyramid was erected to mark the burial of a particular king, and fortunately the builders left inscriptions that attest to the ownership of many of these monuments. The three great pyramids at Giza, for example, were built for rulers named Khufu, Khafre, and Menkaure. Other pyramids were built for kings named Snofru, Sahure, and Neferirkare. Even more fortunately, the ancient Egyptians maintained detailed records of the hundreds of kings who ruled Egypt over the centuries. These lists provide the basis for modern reconstructions of ancient Egyptian chronology and are an essential tool in any estimate of the age of the pyramids.

The most famous of these king lists was produced by an Egyptian priest named Manetho, who lived during the third century BCE. His work includes a list of over two hundred kings who had ruled Egypt since the beginning of recorded history. These kings are grouped together into some thirty dynasties. It is not always clear what all the kings in any given dynasty had in common. Sometimes the kings belonging to a particular dynasty seem to belong to a single family, but in other cases rulers from different families are grouped together. In spite of such uncertainties, Manetho's dynasties still provide the basic nomenclature for organizing Egyptian kings and Egyptian history.

Manetho's annals indicate how long each dynasty lasted and suggest that the total length of time covered by all the dynasties extends over thousands of years. Given such a vast amount of time, there are bound to be issues of accuracy. Furthermore, Manetho's work is preserved only in a few second-hand copies created hundreds of years after he died, and even these are not entirely consistent. Thus while the Manetho tradition is a valuable resource, earlier lists of kings are still needed in order to construct a more accurate chronology.

Luckily for Egyptologists, there are a number of other king lists that are thousands of years older than Manetho's. Some were created to decorate the walls of temples and include only rulers considered "important" at the time the stone was carved, omitting a number of short-lived or heretical kings known from other historical records. Other texts, preserved only in fragments, appear to have originally contained a more or less complete account of Egyptian rulers. For example, the Palermo Stone gives a year-by-year record of events in Egyptian history up through the Fifth Dynasty. There is also the Turin Papyrus, likely created in the Nineteenth Dynasty, which contains a list of kings together with the lengths of their reigns. Here even kings who ruled for only a year are included. Such resources, while regrettably incomplete, still preserve invaluable information that has allowed scholars to come to a clearer understanding of a complex story spanning three millennia.

Based on these and other historical documents, Egyptologists have constructed the outline of ancient Egyptian history shown in Figure 3.1. It is still divided into thirty-one dynasties, which are now grouped together into an Archaic Period, Old, Middle, and New Kingdoms, the Late Period, and three intermediate periods. The kingdoms are, in general, times when the central government of Egypt was strong, while the intermediate periods were times when pharaonic authority was weak. During the latter periods multiple rulers and even foreign powers could exercise control over different parts of Egypt.

The kings associated with the Great Pyramids at Giza belong to the Fourth Dynasty in the Old Kingdom, which places them in a very early period in Egypt's long history. Their pyramids are consequently among the older monuments in Egypt. However, the information from the king lists cannot provide us with a very precise estimate of when these kings lived or when the pyramids were built. These records may indicate how long each king ruled, but there are gaps in the records and, to complicate matters further, some kings may have had overlapping reigns. Uncertainties of this sort tend to accumulate as we go backwards in time, so the timing of very early events, such as the construction of the pyramids, is quite difficult to pinpoint.

It is the intermediate periods that pose the most significant challenges. For the kingdoms, when the central authority of Egypt was strong, the sequence of rulers is fairly certain. The number of years each king ruled during each of

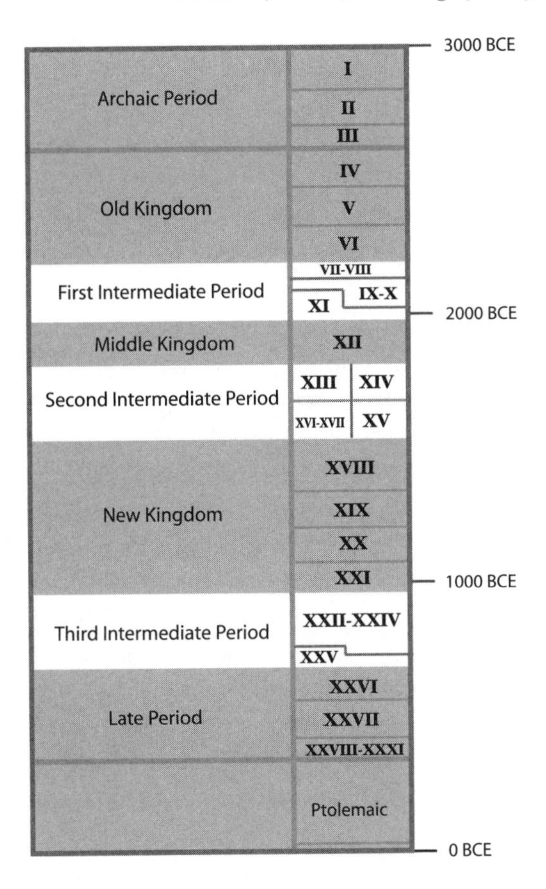

FIGURE 3.1 A rough outline of the history of ancient Egypt. The roman numerals correspond to the dynasties, while the kingdoms and periods are shown at left. The years at the right are only approximate.

these periods is also reasonably well established based on the king lists and contemporary records. By contrast, during the intermediate periods it is not even certain how many rulers there were. Therefore, the kingdoms can float back and forth over several hundred years depending on how long the intermediate periods lasted.

Perhaps surprisingly, astronomical data provide a way to reduce these uncertainties. Documents from the Middle and New Kingdoms record certain astronomical events that allow us to anchor these periods in time. No such document has yet been found from the Old Kingdom itself, but these sources, combined with data encoded in the alignment of the pyramids, may provide a way to estimate the age of the entire Old Kingdom.

SECTION 3.2: OF DOG STARS AND LEAP YEARS: DATING THE MIDDLE KINGDOM

The documents that reveal the age of the Middle and New Kingdoms do not record a supernova, eclipse, or other similarly unique astronomical event, but instead refer to phenomena occurring every year. These events can help shed light on the timing of historical events thanks to a useful quirk of the Egyptian calendar. The calendar used by the Egyptians to regulate administrative affairs consisted of three seasons, each composed of four months of 30 days, plus 5 extra days to make a total of 365 days. This year was in principle tied to seasonal and astronomical events. The names of the three seasons can be loosely translated as "flood," "growing," and "harvest." "Flood" refers to an annual event in which the Nile, due to increased rainfall in the Ethiopian highlands, swells and overflows its banks, depositing a new layer of rich soil over the Egyptian fields. After the flood recedes, the crop is planted and the "growing" season starts. Finally, there is a "harvest" season before the next flood. Of course, the flood could start on different days due to varying weather conditions, so a calendar tied directly to the flood would not be practical. However, an astronomical event always heralded the beginning of this important flood: the "heliacal rising" of Sirius.

Sirius is a very bright star in the constellation of Canis Major. For some time during the year this star disappears behind the sun. After about 70 days, it can again be seen low in the sky just before dawn. This event occurs in July, just about the time of the annual flood. The Egyptians quite reasonably took this event to mark the beginning of the New Year.

However, the Egyptians apparently never used a leap day in their civil calendar, so their year was slightly shorter than the time it takes for the earth to go around the sun, which is about 365.25 days. This means that the time between heliacal risings of Sirius (tied to the astronomical year) was slightly longer than the 365 days of the Egyptian calendar. After four years the rising of Sirius thus occurs on the day before the Egyptian New Year's Day, and as the years go by, this event occurs earlier and earlier in the year. Only after about 1,460 years would Sirius again appear on the "correct" day.

This quirk has proven to be quite useful for historians. Records from Roman times tell us that the first day of the Egyptian calendar and the heliacal rising of Sirius coincided in roughly the year 139 CE. Counting backwards, we can determine that this synchronicity would have also occurred around 1320 BCE and 2780 BCE. We can also calculate when the heliacal rising of Sirius would occur for any other year. Therefore, if a document tells us the rising occurred on a certain day, we can calculate what year is being described.

Fortunately, there are at least two such relics, one from the New Kingdom and one from the Middle Kingdom. Since it is easier to interpret, we will here consider only the Middle Kingdom document, an unimpressive scrap of papyrus known by the equally unimpressive name of "Berlin Museum papyrus 10012." The text on this object records that Sirius would reappear on the sixteenth day of the eighth month in the seventh year of a king named Senusret, which is either 226 days after or 139 days before the Egyptian New Year's Day. This could occur only in or around 1872 BCE, as other possible dates like 412 BCE and 3332 BCE can be easily ruled out based on other archaeological data. Although complications in interpreting this document prevent this date from being nailed down to a single year, it does provide an anchor that fixes the Middle Kingdom in time.

With this information about the age of the Middle Kingdom, the biggest remaining obstacle to determining the age of the Old Kingdom and the pyramids is the First Intermediate Period. Without a similar record of useful astronomical events from the Old Kingdom, the age of the pyramids is uncertain by as much as 200 years. As these monuments are about 4,500 years old, this is only a 5% uncertainty in the age. Nevertheless, even 100 years is more than one generation, and is beyond the living memory of most people, and without a more precise measure of the age of the Old Kingdom, it is very difficult to interpret some of the records of the First Intermediate Period or the early Middle Kingdom, since it is impossible to tell if the last days of the Old Kingdom were considered "ancient history" at this time. This difficulty could be resolved if the uncertainty in the age of the Old Kingdom was reduced to only a decade or two. In 2000, an Egyptologist named K. E. Spence suggested a way to date the Old Kingdom more precisely. It turns out that the pyramids themselves may hold the key to precisely locating the Old Kingdom in time.

SECTION 3.3: THE SEQUENCE OF PYRAMIDS

The pyramids can serve as both monuments and timekeepers because we can use historical records to establish the order in which they were built. While the precursors and prototypes of pyramids were built during the Third Dynasty, pyramid building became a regular industry at the dawn of

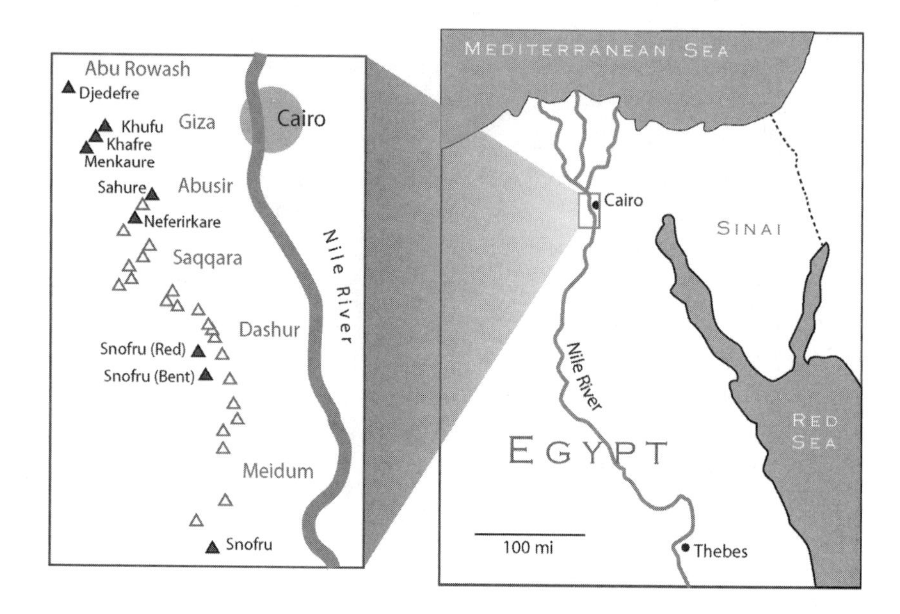

FIGURE 3.2 A map of Egypt, with a close-up showing the locations of the pyramids discussed in the text as solid triangles. The open triangles indicate the locations of other pyramids.

the Fourth Dynasty (Figure 3.2). Snofru, the first ruler of the Fourth Dynasty, was responsible for three large pyramids. The first was built at a site called Meidum, which has partially collapsed and does not look much like a classic pyramid today. Snofru's other two pyramids were built slightly to the north at a place called Dashur. His "Bent" pyramid has a curious shape, since the sides of the pyramid have a steep slope at the base and a shallow slope near the top. This may have been done to reduce the weight bearing down on the inner chambers of the pyramid, whose walls appear to have cracked under stress. Snofru's final pyramid was the "Red" pyramid (named after the color of its limestone blocks), which was built during his third decade on the throne and is probably where he was finally buried.

Snofru's successors were responsible for the famous pyramids at Giza. The first and largest was built for a king named Khufu, who was in all likelihood Snofru's son. Khufu was succeeded by a king named Djedefre, who ruled for about eight years and began a pyramid not at Giza, but at a site called Abu Rowash about ten kilometers to the north. Djedefre died when his pyramid was still only in its early stages, and the site now consists of a few courses of stone and a trench excavated for the burial chamber. Khafre succeeded

Djedefre and was responsible for the other large pyramid at Giza. The third Giza pyramid is smaller than its neighbors and was built by Menkaure, who may or may not have been the direct successor of Khafre. After Menkaure, there were a couple of short-lived kings—who did not construct pyramids as far as we know—who ruled before the end of the Fourth Dynasty.

The details of the transition between the Fourth and Fifth Dynasties are unclear and shrouded in myth. Several rulers in this new dynasty, including Sahure and Neferirkare, built their own pyramids south of Giza, in a place called Abusir. These monuments are a good deal smaller those at Giza. Even so, they must have been impressive structures for many hundreds of years.

Pyramids were built throughout the Fifth Dynasty and into the Sixth, at which time they included texts describing the afterlife of the rulers. These "pyramid texts" form the oldest corpus of religious literature in Egypt. Pyramids were also built in the Middle Kingdom, although by then they were constructed mainly from mud-brick instead of stone and are now badly ruined.

SECTION 3.4: A SUSPICIOUS PATTERN IN THE PYRAMIDS

The pyramids of the early Old Kingdom are impressive feats of engineering not only because of their sheer bulk but also because of their accurate layout. Although much of the outer layers of stone have been stripped away, various marks at the base of the pyramids are enough to document the care with which these structures were designed.

Of particular interest here is that these pyramids were aligned so their sides face the cardinal directions with errors of less than one degree, and the two large pyramids at Giza have errors of less than one tenth of one degree. This degree of accuracy could be achieved only by using astronomical measurements, but which astronomical object did the Egyptians use? Unfortunately, no ancient records explicitly document the procedures used in surveying and laying out pyramids, and there are many possible ways the Egyptians could have determined the proper orientation of the monuments. Most of the methods considered by Egyptologists would—if done properly—always yield a structure perfectly aligned with the cardinal directions. For example, by tracking the motion of a star across the sky or the movement of a shadow along level ground, the Egyptians could have figured out when the sun or some star reached the highest point in the sky, at which point it must be either due north or due south. While such methods do have an appealing elegance to them, there is evidence that the Egyptians did not use such a reliable approach.

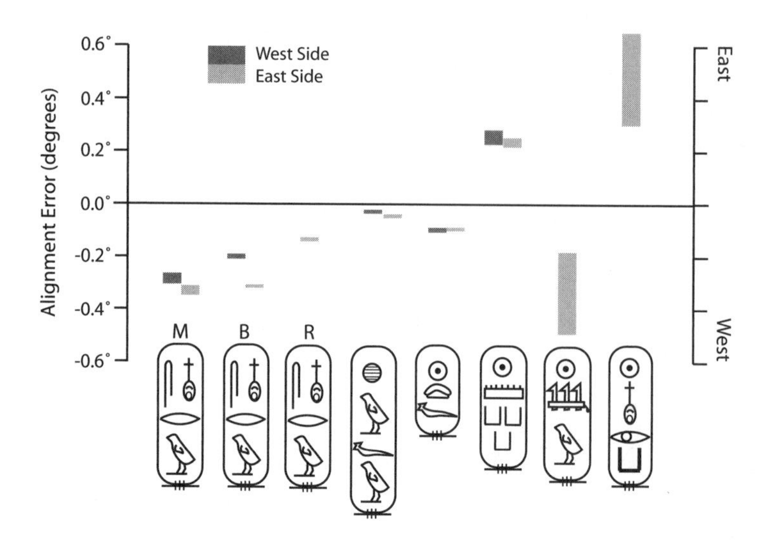

FIGURE 3.3 The errors in the alignments of the pyramids in the order of their construction (data from Kate Spence "Ancient Egyptian Chronology and the Astronomical Orientation of the Pyramids" *Nature* 408 (2000): 320–24, p. 320). The relevant king names are given at the bottom (from left to right they are Snofru, Khufu, Khafre, Menkaure, Sahure, and Neferirkare). The letters above Snofru's name refer to his three pyramids (Meidum, Bent, and Red). For each pyramid, bars indicate the alignment error of the east and west sides of particular pyramid (the height of each bar shows the uncertainty of the measurement). Note that in general the pyramids start out skewed slightly to the west but with time gradually come to skew eastwards.

If one takes the pyramids in the order they were built and plots the deviation in their orientation from true north (which I call the alignment error) for each one, then we get the intriguing pattern shown in Figure 3.3. The large pyramids at Giza built by Khufu and Khafre have the smallest alignment errors and are aligned most closely with true north, while the pyramids built before and after this time have noticeably larger alignment errors. What is particularly interesting, however, is that all of Snofru's pyramids, built before the Giza pyramids, are skewed slightly to the west, while two out of three of the later pyramids are skewed slightly to the east.

This pattern becomes even more suspicious when we include information about when the pyramids were built relative to each other. Recall that while the historical records alone cannot indicate how long ago the pyramids were built, they do tell us how long each of the kings ruled. For example, the Turin Papyrus records that Khufu ruled for twenty-three years and Djedefre ruled for eight years, so a total of thirty-one years separates Khufu's and Khafre's ascensions to the throne of Egypt. Since the construction of a pyramid is a major project requiring a decade or two to complete, and it had to be finished in order for the king to have a proper burial, it is reasonable to expect that the work on the monument for a particular king would start soon after the coronation. Therefore, we can estimate that Khafre's pyramid was laid out some thirty years after Khufu's. With other historical records, we can also calculate how many years separate the other pyramids from Khafre's pyramid. Then we can make the plot shown in Figure 3.4, which shows the alignment error of each pyramid as a function of when that pyramid was laid out relative to Khafre's (or any other king's). Now we can see that the data for most of the pyramids seem to fit to a line, and the two pyramids that fall significantly far away from this line—Khafre's and Sahure's—would fall on it if the sign of their alignment errors are changed from negative (westward) to positive (eastward). We will see soon that flipping the sign of the alignment error may be a reasonable thing to do.

This pattern might be just an interesting coincidence. However, the slope of this line indicates that these pyramids' orientation was changing at a steady rate of about half a degree per century. This drift rate is significant because it means the alignment errors could be changing with time because the technique used to lay out the pyramids was affected by the precession of the Earth.

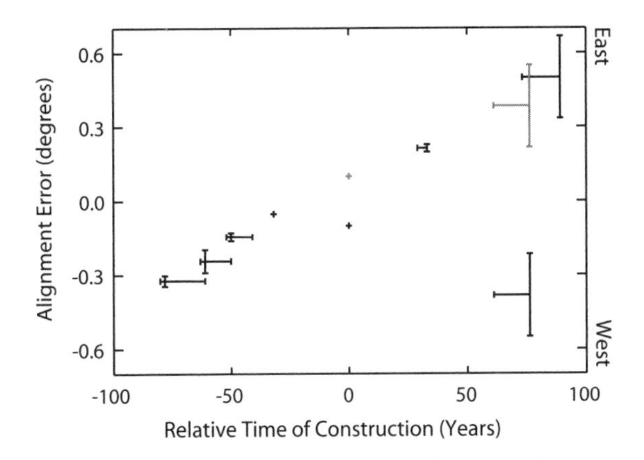

FIGURE 3.4 The alignment errors in the pyramids as a function of time (data from Kate Spence "Ancient Egyptian Chronology and the Astronomical Orientation of the Pyramids" *Nature* 408 (2000): 320). The black crosses give the measured values and uncertainties of the alignment errors along the y-axis and the times when the pyramids were built relative to each other along the x-axis. Flipping the sign of the alignment error of the two pyramids that don't follow the eastward trend gives the gray data points.

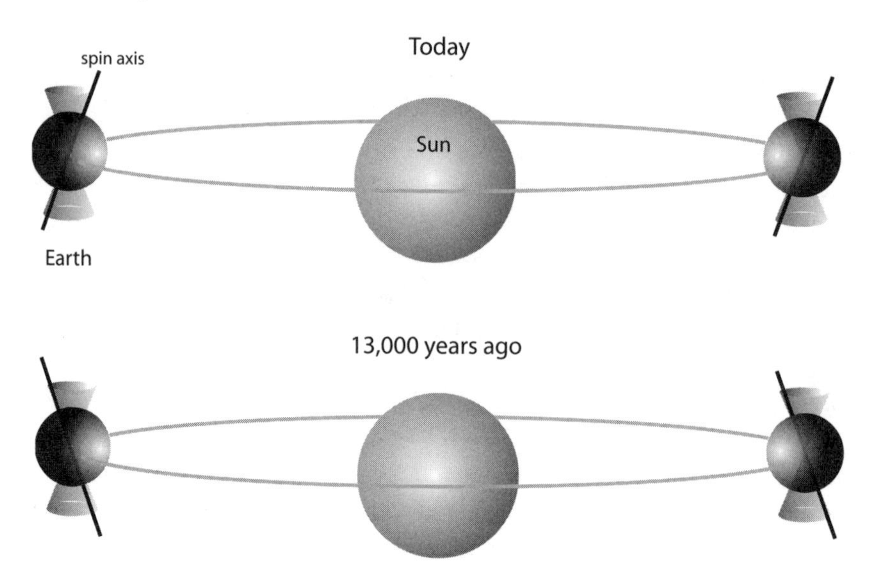

FIGURE 3.5 Precession of the earth. The top panel shows the earth in two positions in its orbit around the sun (not to scale). The spin axis is indicated by the bar going through the earth, which points in the same direction in space as the earth moves around the sun. As the earth precesses, this axis changes direction and traces out a circle. Thirteen thousand years ago the axis of the earth pointed in another direction, as shown in the bottom panel.

SECTION 3.5: SPINNING TOPS AND TWISTING PYRAMIDS

Precession is a phenomenon that occurs in spinning bodies under the influence of asymmetric forces. A familiar example of precession occurs with tops and gyroscopes. A top that is not spinning will just fall on its side, but if the top is spinning fast enough, it manages to keep from toppling over. Even so, the force of gravity still has an effect on the top. If there were no outside forces or if the top were freely falling, it would spin about its axis and that axis would always point in the same direction. However, when a table or some other surface supports the point of the top, the asymmetric distribution of forces on the top cause the spin axis to change direction with time. In particular, the free end of the axis of the top traces out a horizontal circle. This motion of the spin axis is called the *precession* of the top.

The earth rotates once a day on an axis that today points at the star called Polaris. This axis continues to point in this same direction even as the earth moves around the sun, as shown in Figure 3.5. This is why Polaris appears to the north throughout the year. However, the spin axis does slowly precess due to the gravitational forces from the sun and the moon acting on the slightly oblate earth. Just as the spin axis of a top traces out a horizontal circle, the spin axis of the earth traces out a circle that is parallel to the plane of earth's orbit.

For those of us living on earth, this precession affects the appearance of the night sky. The stars at night appear to circle around a point defined by the earth's spin axis called the celestial pole. Right now, the celestial pole is near Polaris, the "pole star." However, as the earth precesses the celestial pole moves through the stars in a wide circle centered on the constellation of Draco (shown in Figure 3.6), so thousands of years ago Polaris was not near the celestial pole and the northern sky looked quite different than it does today (see Figure 3.7). This movement of the celestial pole is determined by the shapes, spins, masses, and orbits of the sun, the moon, and the earth, and thus we can calculate where the celestial pole was relative to the stars at any given time in the past.

Since the mid-1980s some researchers have hypothesized that the changes in the alignment of the pyramids could be related to the precession of the

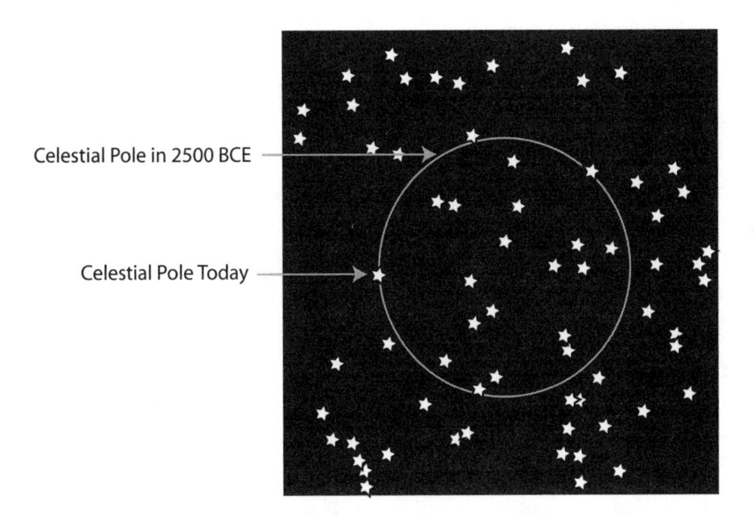

FIGURE 3.6 The path of the celestial pole. Due to the precession of the earth, the celestial pole (the point the stars appear to move around) moves around the marked circle once every 26,000 years. For scale, note the Big Dipper in the upper left of the drawing.

FIGURE 3.7 The changing night sky. On the left we see the night sky as it is now. All stars appear to orbit Polaris (near the center of the image). On the right we have the sky as it appeared when the pyramids were built. Note that the stars have changed position, and no bright star was exactly where Polaris is today.

earth. If true, this would provide important information about the procedures the Egyptians used to lay out these structures. Most of the methods that have been proposed for aligning the pyramids are insensitive to the precession of the earth. For example, it has often been suggested that the Egyptians determined the location of the North Pole by observing where a star rose and set along a level surface. A point midway between these two locations is due north (or south) of the observer regardless of where the star lies in the sky, so this method always gives a reliable estimate of true north. This and similar methods involving the sun therefore cannot explain why the alignment of the pyramids drifts over time.

Prior to K. E. Spence's paper, some scholars had proposed ways to align the pyramids using the positions of certain stars to *approximate* the location of the celestial pole, so that the alignment of the pyramids would change as the celestial pole moves slowly through the sky. Spence's work, however, sparked a renewed interest in these ideas because she proposed that the drift of the alignment error could not only provide evidence that the Egyptians used stars to align the pyramids, it could also tell us which stars they used, and in turn reveal the exact dates when the pyramids were built.

SECTION 3.6: A NOVEL METHOD FOR ALIGNING THE PYRAMIDS

During the era of pyramid construction the celestial pole was not near any particularly bright star, so there was no analog of Polaris to indicate the direction of true north. Without a pole star available to them, the Egyptians must have used the stars in some other way to align the pyramids. Whatever this method was, it should depend upon the location of the stars with respect to the celestial pole, so that the precession of the earth would cause the alignment of the pyramids to drift with time.

Spence suggested that the Egyptians used two stars in the northern sky to "point" to the celestial pole and find north. She supposes that when the pyramids were built, there were two stars positioned in the sky such that the line connecting them passed close to the celestial pole. If there is such a pair of stars, then when the stars were aligned vertically, a vertical line connecting the stars and extended to the ground (i.e., a plumb bob) indicated the direction of true north, as shown in Figure 3.8.

FIGURE 3.8 Using two stars to align the pyramid. If the correct pair of stars is chosen (highlighted with the circle and a square), then extrapolating a vertical line connecting the two stars to the ground (for example, with a plumb bob) can provide a good indication of north.

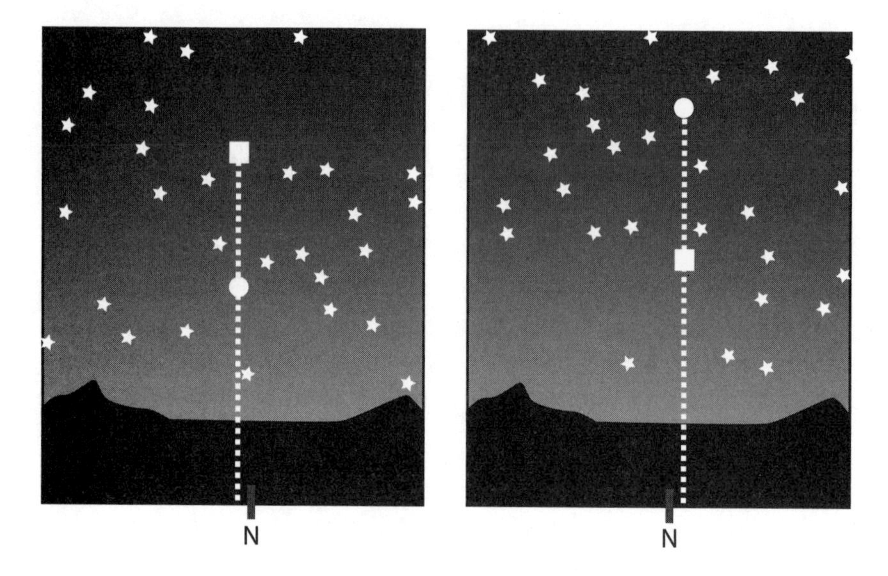

FIGURE 3.9 Different errors with the same stars. The same pair of stars (indicated with a circle and a square) is used to determine north, but a different star is higher in the sky in the two plots. The line connecting the stars is used to estimate north (indicated by the short bar at the bottom of the picture). In both cases, there is an error in the measure of north, but in one case the orientation is slightly to the left (west), and in the other it is skewed to the right (east).

This suggestion is certainly plausible, and if correct the alignment of the pyramids will drift with respect to true north as the celestial pole moves with respect to the line joining the stars. Furthermore, it naturally explains why two of the pyramids seem to have alignment errors with the "wrong sign," that is, skewed west instead of east. The sign of the alignment error depends on which of the two stars was higher in the sky when the measurement was made (see Figure 3.9). This suggests that six of the pyramids were aligned with one star higher in the sky, and the other two were aligned with the other star higher in the sky, perhaps because they were laid out during a different time of the year.

Another attractive feature of this method is that it uses northern stars, which we know were very important to the Egyptians. They are mentioned in the pyramid texts inscribed on the later Old Kingdom pyramids, which describe them as the "indestructible stars" because they never went below the horizon and so never entered the underworld. These texts also indicate that the rulers of Egypt wanted to be identified with these immortal stars, so it would make sense that they would use them in aligning their pyramids.

SECTION 3.7: DID THE EGYPTIANS REALLY DO THIS?

If the pyramids were really aligned as Spence proposes, and the changing alignment of the pyramids tracks a predictable astronomical phenomenon, then it follows that the varying orientations of the pyramids could indicate their age. Of course, we need to know which star pair the Egyptians actually used and, more fundamentally, we must have evidence that Egyptians used this method and not some other technique. What makes Spence's proposal especially interesting is that the alignment data can themselves be used to verify that the Egyptians used this method. Furthermore, we can establish the particular pair of stars the Egyptians observed without knowing how the Egyptians made the measurement or the religious significance of different stars in the northern sky.

At any given time, the celestial pole moves through the sky in a particular direction. The rate at which the alignment error changes with time depends on how the line connecting the stars is oriented with respect to this direction of motion. If the path of the celestial pole is almost perpendicular to the line connecting the stars, then the celestial pole spends comparatively little time near the line and the alignment error will change relatively rapidly with time. On the other hand, if the path of the celestial pole passes through the line joining the stars at a shallow angle, then the celestial pole spends a longer time near the line and the alignment error will change more slowly. Therefore, we can compare the expected drift rate in the alignment derived from different pairs of stars to the rate observed in the pyramids and find the star pair that gives the best match, this being the one the Egyptians most likely used. If it turns out that no pair of stars provides a rate close to that observed in the pyramids, then the Egyptians must have used another method. This makes this proposal particularly attractive because instead of relying upon indirect arguments regarding the ancients' abilities or interests, we can use the orientation data directly to establish if a given method was used to orient the pyramids.

Spence did not do an exhaustive survey of all possible star pairs, but such a search can be easily done. First, to narrow down the list of potential star pairs, we will consider only the brightest stars within 25 degrees of the celestial pole in the year 2500 BCE, approximately when the Great Pyramids were built. There are 17 stars that meet these criteria, which makes for 136 possible pairs. For each pair, it is straightforward to calculate approximately when a line

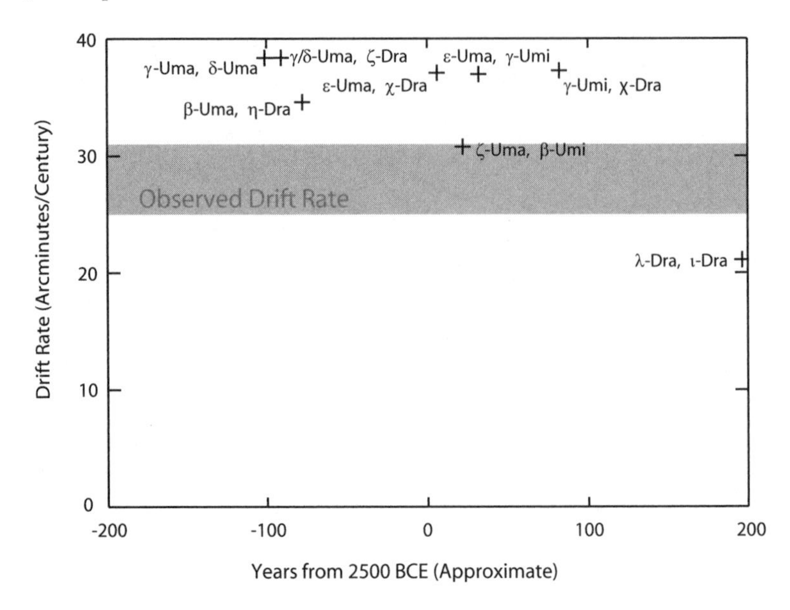

FIGURE 3.10 For every possible pair of bright stars in the northern sky, we compute the rate at which the alignment error should change with time and the date when the measurement should give exactly true north. These two variables are plotted here, the drift rate given in units of arcminutes (60 arcminutes = 1 degree) per century. Only nine pairs actually appear in the time range shown here, and only one fits in the range consistent with the observed drift rate (indicated by the shaded bar). If the Egyptians did indeed employ this method, this is most likely the pair that they used.

passing through both stars passes through the celestial pole and how fast the alignment error would change with time (see the appendix for details). Only nine pairs yield lines that go near the celestial pole within 200 years of 2500 BCE, so these are the most reasonable candidates. As shown in Figure 3.10, the predicted rate of change in the alignment errors for most of these pairs are around 35–40 arcminutes (about two-thirds of a degree) per century, one is about 30 arcminutes (about one-half of a degree) per century, and one is about 20 arcminutes (about one-third of a degree) per century.

Do any of these rates match the rate observed in the pyramids? Given the best estimates of the orientations of the pyramids and the timing of their construction, we obtain a rate of 28 arcminutes per century, with an uncertainty of about 3 arcminutes per century (see the appendix for details of this calculation). Amazingly, only one of the above pairs of stars is consistent with this rate. This pair is Mizar (in the Big Dipper) and Kochab (in the Little Dipper), which are also known as ζ Ursa Majoris and β Ursa Minoris, the stars in the dark bar in Figure 3.10. If the Egyptians did orient the pyramids as Spence proposes, these stars are the only pair of stars that they could have used.

Since the observed changes in the alignment of the pyramids are consistent with one and only one pair of stars, and we know when the line connecting those stars would give any of the observed alignment errors, we can ascribe unambiguous dates to all of these pyramids (assuming the ancient Egyptians used this method). For example, the Great Pyramid of Khufu at Giza would appear to have been laid out in 2480 BCE, give or take a few years. This date is fifty years later than the current best guess based on historical records, but it is also not unreasonable.

SECTION 3.8: CONFIRMING THE METHOD: ET TU, DJEDEFRE?

The above method of dating the Old Kingdom is certainly intriguing, but additional data are clearly needed to verify that the pyramids were aligned using the stars as Spence suggests. These data could come from refined measurements of the orientation and relative dates of the pyramids, or from measurements of the orientations of the incomplete pyramids started by various minor rulers of this period, such as the unfinished pyramid of Djedefre.

As mentioned above, Djedefre ruled briefly between Khufu and Khafre. He began a pyramid in Abu Rowash, north of Giza, but construction had only just begun before it was abandoned—presumably due to the ruler's untimely death. Since Djedefre's pyramid was laid out when the line connecting Kochab and Mizar went almost exactly through the celestial pole, this unfinished monument should be the most accurately aligned "pyramid" of all. In 2003, a French team published the orientation data of the pyramid, and found that it was skewed west of north by 0.8 degrees. It is actually *less* well aligned than any other pyramid from the era.

This aberrant monument thus fails to support Spence's theory of pyramid surveying. However, it does not decisively falsify it, either. Djedefre chose to be buried far from Giza and the rest of his family, and his successors chose not to follow his example, so one might suggest he also elected to use a different alignment method.

Personally, I think a better approach in testing Spence's idea is to improve the orientation data of the pyramids that appear to follow the predicted trend.

At present, the orientations of the Great Pyramids of Giza are very well measured to within less than an arcminute. However, the measured orientations of the other pyramids are uncertain by as much as 10 arcminutes (see appendix). If these uncertainties could be reduced to levels comparable to those of the Great Pyramids, then we can see if the Fifth Dynasty pyramids of Sahure and Neferirkare really follow the same trend as the Great Pyramids. If they do not, then there is much less support for this model. Furthermore, if the various monuments still fall along the same line, then the uncertainty in the drift rate will be reduced. If this rate remains consistent with the drift in the positions of Mizar and Kochab, then this new data would provide strong support for the model. On the other hand, if the refined estimate of the observed drift rate is inconsistent with the movements of these stars, then the Egyptians most likely did not use this method for aligning the pyramids.

Even if this method of determining the age of the pyramids turns out to be wrong, it is still an interesting example of how historians attempt to measure age in the absence of complete calendrical information. Clearly, combining historical, archaeological, and astronomical data occasionally allows us to figure out when things happened with great precision, and with residual uncertainties as small as a few years. However, these streams of information can be combined only in rather extraordinary situations, such as those involving gigantic monuments aligned with great precision. Different techniques are therefore needed to date the vast majority of the relics from the past.

SECTION 3.9: FURTHER READING

Some good on-line resources on ancient Egypt in general and the pyramids in particular are the Giza Archives project, www.gizapyramids.org; www.egyptology.com, which contains many good links; and the Abzu archive www.etana.org/abzu, which links to many scholarly articles about Egypt and the ancient Near East.

For someone completely unfamiliar with Egyptian history, I recommend starting with Barbara Mertz *Temples, Tombs, and Hieroglyphics* (Bedrick, 1990). For more detailed general works on Egyptian history, see Peter A. Clayton *Chronicle of the Pharaohs* (Thames and Hudson, 1994) and Ian Shaw *The Oxford History of Ancient Egypt* (Oxford University Press, 2000).

A good general work on the pyramids is Miroslav Verner *The Pyramids* (Grove Press, 2001). For more information on how they may have been built, try Dieter Arnold *Building in Egypt* (Oxford University Press, 1991) and Martin Isler *Sticks, Stones, and Shadows* (University of Oklahoma Press, 2001).

An excellent popular work explaining the basic mechanics of precession, etc., is Larry Gonick and Art Huffman *Cartoon Guide to Physics* (Harper Perennial, 1991).

The classic reference work on Egyptian calendars is Richard A. Parker *The Calendars of Ancient Egypt* (University of Chicago Press, 1950). For more recent efforts to refine the dating of the Middle Kingdom, try the references in Leo Depuydt "Sothic Chronology in the Old Kingdom" *Journal of the American Research Center in Egypt* 37 (2000): 167–186.

The original article on using precession to date the pyramids is Kate Spence "Ancient Egyptian Chronology and the Astronomical Orientation of the Pyramids" in *Nature* 408 (2000): 320–324. A spirited exchange of letters responding to this article and pointing out a math error in the original analysis are found in *Nature* 412 (2001): 699–700.

A prior article that also noted precession as an explanation to the pyramid alignment errors is Steven C. Haack "The Astronomical Orientation of the Pyramids" *Archaeoastronomy*, no. 7 (1984): S119–S125.

Juan Antonio Belmonte "On the Orientation of the Old Kingdom Egyptian Pyramids" *Archaeoastronomy*, no. 26 (2001): S1–S20, follows on Spence's work and suggests another pair of stars (which are inconsistent with the observed drift rate). It also gives a good discussion of issues involved in stellar alignments.

The new measurements of Djedefre's pyramid can be found in Bernard Mathieu, "Travaux de l'IFAO en 2000–2001" *BIFAO (Bulletin de l'Institut francais d'archeologie orientale)* 102 (2002): 437–614, at 458.

APPENDIX: THE DRIFT RATE CALCULATIONS

The drift rate calculations illustrated in Figure 3.10 cannot be found in the published literature. Therefore I am including here the details of the calculations used to make this plot.

Only stars brighter than magnitude 4 and within 25 degrees of the celestial pole in 2500 BCE are considered as reasonable candidates for aligning the pyramids. The 17 stars that meet these criteria are α, β, γ, δ, ε, ζ, and η Ursa Majoris; α, β, and γ Ursa Minoris; and α, ζ, η, ι, κ, λ, and χ Draconis. The positions of these stars are given in standard bright star catalogs, and for these calculations the proper motions of these stars are neglected.

After computing the locations of these stars relative to the position of the celestial pole in 2500 BCE, a line is drawn through each pair of stars. This line is projected through a straight line approximating the path of the celestial pole around 2500 BCE (the curvature of this path can be neglected for the range of time considered here). The point of intersection between these two lines indicates when the celestial pole fell along the line joining the two stars, and the angle between the lines gives the drift rate.

As for the observed drift rate in the orientation of the pyramids, the calculation starts with the orientations of the pyramids given in Spence's article (see Table 3.1). The effective alignment error in the last column is the weighted average of the two sides, and the sign of the errors in Khafre's and Sahure's pyramids has been flipped as discussed above. The uncertainties are the larger of the measured uncertainty and 1/2 the difference between the orientations of the two sides. Only one side of the Red Pyramid is measured, so the error on this pyramid is set rather arbitrarily to 1′ (changing the value does not affect the results significantly).

TABLE 3.1 Alignment errors in pyramids (in arcminutes = 1/60 of a degree).

Pyramid	Alignment Error (West Side)	Alignment Error (East Side)	Effective Alignment Error
Snofru (Meidum)	-18.1' ± 1.0'	-20.6' ± 1.0'	-19.4' ± 1.3'
Snofru (Bent)	-11.8' ± 0.2'	-17.3' ± 0.2'	-14.6' ± 2.8'
Snofru (Red)	—	-8.7' ± 0.2'	-8.7' ± 1.0'
Khufu	-2.8' ± 0.2'	-3.4' ± 0.2'	-3.1' ± 0.3'
Khafre	-6.0' ± 0.2'	-6.0' ± 0.2'	+6.0' ± 0.3'
Menkaure	+14.1' ± 1.8'	+12.4' ± 1.0'	+12.8' ± 0.9'
Sahure	—	-23' ± 10'	+23' ± 10'
Neferirkare	—	+30' ± 10'	+30' ± 10'

TABLE 3.2 Alignment change between chronologically sequential pyramids (in arcminutes = 1/60 of a degree).

Pyramid Pair	Orientation Change
Snofru (Meidum)–Snofru (Bent)	4.8' ± 3.1'
Snofru (Bent)–Snofru (Red)	5.9' ± 3.0'
Snofru (Red)–Khufu	5.6' ± 1.0'
Khufu–Khafre	9.1' ± 0.4'
Khafre–Menkaure	6.8' ± 1.0'
Menkaure–Sahure	10' ± 10'
Sahure–Neferirkare	7' ± 14'

TABLE 3.3 Elapsed time between construction of chronologically sequential pyramids. Data from Kate Spence "Ancient Egyptian Chronology and the Astronomical Orientation of the Pyramids" *Nature* 408 (2000): 320–324.

Pyramid Pair	Elapsed Time (years)
Snofru (Meidum)–Snofru (Bent)	12–17
Snofru (Bent)–Snofru (Red)	9–11
Snofru (Red)–Khufu	9–20
Khufu–Khafre	31–32
Khafre–Menkaure	30–33
Menkaure–Sahure	32–43
Sahure–Neferirkare	12–13

TABLE 3.4 Alignment change between chronologically sequential pyramids.

Pyramid Pair	Rate (arcminutes/century)
Snofru (Meidum)–Snofru (Bent)	33 ± 22
Snofru (Bent)–Snofru (Red)	59 ± 30
Snofru (Red)–Khufu	39 ± 16
Khufu–Khafre	29 ± 1.5
Khafre–Menkaure	22 ± 3
Menkaure–Sahure	27 ± 27
Sahure–Neferirkare	56 ± 113

Based on these data we can calculate how much the alignment changed between sequential pairs of pyramids (see Table 3.2).

The amount of time that elapsed between the construction of these pyramids is then obtained from historical records (Table 3.3). Taking the uncertainty in the timing as 1/2 of this range of years, we can calculate the rate at which the alignment changed for each pair (Table 3.4).

Combining all these estimates, we get the final rate of 28 arcminutes per century and a 2-sigma range of 3 arcminutes per century. This range is somewhat conservative, and a more refined estimate could be made by quantifying the various uncertainties more carefully, but such calculations are beyond the scope of this book.

CHAPTER FOUR

The Physics of Carbon-14

The age of an artifact does not have to be inscribed on its surface or encoded in its orientation. In certain cases, just a sample of atoms from an object can reveal how old it is. For example, measurements of the number of carbon-14 atoms preserved in material like wood or bone provide many of the best dates for archaeological sites. Since these carbon-14 dates do not require historical documents, they can be used on objects thousands of years older than the earliest written records, enabling archaeologists to explore such topics as the origin of agriculture and the earliest inhabitants of the New World. At the same time, carbon-14 data have helped scientists better understand shifts in earth's climate over the millennia, and have even revealed long-term changes in the surface of the sun.

The tools and procedures used to extract chronological data from atoms like carbon-14 began to be developed in the 1940s and have continued to be improved and refined until the present day. Indeed, as we will see in the following chapters, archaeologists and other researchers still often have interesting and lively debates about how to best obtain and interpret carbon-14 dates. However, in spite of lingering controversies about the details of the method and the interpretation of specific dates, the basic principles and physics behind carbon-14 dating have not changed over the decades. Perhaps because of this, the physics behind carbon-14 dates is rarely discussed in much detail anymore, and phenomena like nuclear decay are often treated in a rather perfunctory way. This is unfortunate, because a closer look reveals that much of the power and limitations of carbon-14 dating derives from fundamental principles of modern physics. In fact, this method of dating archaeological remains involves both the bizarre world of quantum mechanics and Einstein's famous equation $E = mc^2$.

SECTION 4.1: THE HALF-LIVES OF ATOMIC NUCLEI

The basic reason carbon-14 dating is so powerful is that it exploits processes happening deep inside individual atoms. All atoms are composed of a compact nucleus of positively charged protons and neutral neutrons, surrounded by a relatively diffuse cloud of negatively charged electrons. Atoms can exchange or share electrons rather easily, so the chemical properties of an atom depend mainly on how many electrons it has. By contrast, the nucleus is buried so deep in the atom that it does not directly participate in interactions with other atoms except in very extreme circumstances. This means that the state of the nucleus is not usually affected by its chemical environment, and the intrinsic properties of an atom are ultimately established by the number of protons and neutrons it has.

The positive charge of the nucleus is responsible for holding the negatively charged electrons in place, so the number of protons determines the configuration of electrons connected with the atom in a particular situation. Atoms with the same number of protons have the same chemistry and therefore are classified as the same element. Atoms with six protons are all forms of the element carbon, atoms with seven protons are all forms of nitrogen, and so on.

Atoms with the same number of protons can have different numbers of neutrons in their nuclei. Since neutrons are neutral, they do not change the charge of the nucleus and have little influence on the chemistry of the atom. Atoms that have the same number of protons but different numbers of neutrons are called *isotopes* of an element. For example, most carbon has 6 protons and 6 neutrons and is called carbon-12, while carbon-14 is an isotope with 6 protons and 8 neutrons. Different isotopes of the same element can be distinguished from one another because atoms with more neutrons have larger masses. For example, carbon-14 is more massive than carbon-12.

The configuration of protons and neutrons in an atom also determines the stability of the nucleus. Many isotopes, like carbon-12, are completely stable, meaning that they never change into another isotope or element on their own. Others, like carbon-14, can spontaneously transform—or decay—into another type of atom. These transformations are mediated by some of the same forces that hold the nucleus together. Various experiments have shown that these nuclear forces operate over an extremely short range, so that two nuclei must almost touch before they will interact via these forces. On earth, nuclei almost never get this close because they are all positively charged and thus repel each other. This means that how and when unstable nuclei decay is almost completely independent of their environmental conditions, which makes them particularly useful for measuring the passage of time.

Even though the stability and the lifetime of a nucleus depend only on the number of protons and neutrons it contains, the rules that determine if and how a given nucleus transforms are not straightforward. For example, while having more neutrons usually amplifies the forces holding a nucleus together, the two extra neutrons in the carbon-14 nucleus actually make it less stable than ordinary carbon-12. There is a complicated model that describes the relevant interactions between the particles in the nucleus, but we do not need to know all these gory details to establish that carbon-12 is stable and carbon-14 is not. Instead, we can simply use Einstein's famous equation, $E = mc^2$.

This formula encodes one of the most important concepts of modern physics. It tells us that there is a well-defined energy E associated with any massive object, and that the energy is simply the mass m times a factor of c^2, a constant equal to the speed of light squared. This relationship between mass and energy was a revolutionary concept because in classical pre-Einstein physics, mass and energy are very different things. Mass is a quantity intrinsic to an object that determines how it responds to outside forces. Given the same push, an object with less mass will move more rapidly than an object with more mass. Energy, on the other hand, is not necessarily intrinsic to an object—it can change form and can be transferred from one system to another. The defining characteristic of energy is that it is conserved, which means it can neither be created nor destroyed. One basic form of energy is kinetic energy, or the energy of motion: the faster something moves the more kinetic energy it has. The mutable nature of energy means that all other forms of energy can be interpreted as the potential to create motion. The equation $E = mc^2$ posits that any massive object contains a reservoir of "mass-energy," which can be converted into other forms of energy if the mass of the object can be changed.

Such changes in mass play an important part of the physics of atomic nuclei. For example, an ordinary carbon-12 nucleus contains 6 protons and 6 neutrons, but the mass of the nucleus is about one percent smaller than the combined mass of the 12 individual particles. While this discrepancy is difficult to understand in classical terms, where mass is an intrinsic quality of matter, from Einstein's perspective it simply means that the nucleus has less mass-energy than its component parts. We therefore need to add energy to

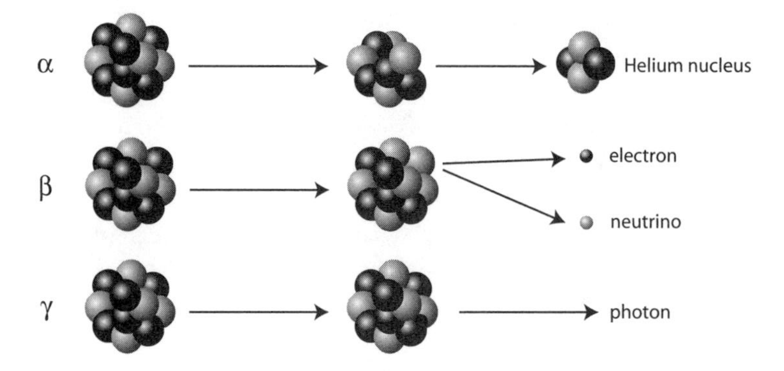

FIGURE 4.1 The three types of nuclear decay. On the left, we have nuclei, made up of protons (gray circles) and neutrons (black circles). These nuclei decay in different ways. At the top we have alpha decay, where the nucleus breaks into two pieces, one of which usually consists of two protons and two neutrons—a helium nucleus. In the middle we have beta decay, where a neutron turns into a proton and emits an electron and a neutrino. At the bottom we have gamma decay, where the nucleus emits a photon.

the nucleus in order to break it into a dozen isolated protons and neutrons. Without this additional energy, the nucleus will never fall apart in this way, and since typical interactions between the nucleus and its electrons or other atoms cannot supply anywhere near enough energy to split the nucleus apart, this process almost never occurs on earth. It turns out that the mass of a carbon-12 nucleus is less than the mass of any other combination of 6 protons and 6 neutrons, so there is no way that the nucleus can break into pieces without some outside source of energy.

However, in the real world a nucleus does not have to simply break into pieces. In fact, there are three different ways nuclei can decay, which are illustrated in Figure 4.1. The most straightforward process is known as alpha decay, which involves the nucleus splitting into two parts. This decay typically yields a nucleus of helium-4, with two protons and two neutrons. There is also beta decay, which occurs when a neutron in the nucleus converts into a proton. This process requires the emission of an electron and a neutrino—a neutral, nearly massless particle. There are variations on this process as well, such as when a proton converts into a neutron, and the nuclear physics behind them all is basically the same. Gamma decay, on the other hand, occurs when the nucleus emits a photon, a particle of light. Other sorts of transformations—such as a neutron simply disappearing from a nucleus—have never been observed, and only alpha and beta types of decay can alter the number of protons or neutrons in the nucleus. This means that only these two processes change one element into another.

Carbon-12 is stable because it cannot undergo either alpha decay or beta decay spontaneously. A carbon-12 nucleus is less massive than any other combination of 6 protons and 6 neutrons, so if it underwent alpha decay, the mass-energy of the nucleus would increase. Similarly, the nucleus would have to gain mass-energy after undergoing a beta decay because carbon-12 is slightly less massive than an atom with 7 protons and 5 neutrons. Both of these processes therefore require some external source of energy to occur and therefore never happen except under special conditions.

By contrast, carbon-14 is unstable and can decay spontaneously. Carbon-14 does not undergo alpha decay, because like carbon-12, it is less massive than the two nuclei produced by such a transformation. However, beta decay is possible. A carbon-14 nucleus contains 6 protons and 8 neutrons, and if one of the neutrons converts into a proton, then we have a common isotope of nitrogen, nitrogen-14, which has 7 protons and 7 neutrons. Nitrogen-14 is less massive than carbon-14—by about one part in 100,000—so no energy needs to be supplied to the nucleus for the transformation to happen. This transformation therefore can and does happen spontaneously, with most of the excess mass-energy converted into the motion of the electron and neutrino emitted by the decaying neutron.

Since the decay of carbon-14 involves only the nucleus itself, the amount of time it takes for the nucleus to change into nitrogen-14 is independent of its surroundings. Again, this is why carbon-14 and other unstable nuclei are such powerful timekeepers. Yet while it is true that the number of protons and neutrons control how long a nucleus will take to decay, this does not mean that all nuclei with the same number of protons and neutrons last the same amount of time. For example, even though all carbon-14 nuclei contain the same number of particles, they do not all decay after, say, 6,000 years. Some carbon-14 nuclei only last a few years, and others survive for tens of thousands of years.

In spite of these variations in the lifetime of individual nuclei, groups of unstable atoms do decay in a predictable fashion with a characteristic time scale known as the half-life. For carbon-14 the half-life is about 5,700 years. This means that if we have a sample of pure carbon-14, then after 5,700 years (one half-life), half of the carbon 14 atoms will have decayed into nitrogen. After another 5,700 years, half of the remaining half will have decayed, leaving a quarter of the original nuclei behind, and so on and so forth. This highly regular behavior, which is illustrated in Figure 4.2, is very useful for dating purposes. This pattern also seems so simple that it is easy to take it for granted, but, in fact, this is a very bizarre phenomenon.

The forces that glue nuclei together are very strong and operate over an extremely limited range, so we expect that the transformation of any given nucleus will not depend in any way on its surroundings. These transformations should therefore be a collection of independent events. Somehow the observed regular behavior emerges from these events, and it is not at all obvious how this might occur. Certainly, there are many situations where a series of independent events produce simple results. Every flip of a coin is an independent event, but after many coin flips, a pattern emerges: heads comes up half of the time. However, physicists strongly suspect that something more than random chance must be influencing how carbon-14 decays, because the curve shown in Figure 4.2 has a specific shape that appears in other systems and often reflects a fundamental aspect of the physics involved.

The implications of this curve's shape are best demonstrated with a more prosaic example. Imagine we have a water-filled chamber with a narrow nozzle at its base, as shown in Figure 4.3. The total weight of the water in the chamber pushes it through the nozzle, causing water to spew out rapidly. As the water level in the chamber falls, there is less pressure at the nozzle, and the flow of water from the tank slows. Suppose the chamber starts out filled with one liter of water and after one second half a liter of water has leaked out of the chamber. The chamber is then half full after one second. Since the amount of water in the chamber has been divided in half, the pressure forcing water through the nozzle is also halved, which means the leak rate will be half what it was originally. The amount of water leaving the chamber in the next second is consequently half that of the previous second: one-quarter of a liter. This leaves one-quarter of a liter in the chamber, or half the volume that was there in the previous second. As this process continues, the amount of water in the chamber is cut is half every second. The volume of water in the chamber therefore has a "half-life" of one second.

This system demonstrates a half-life because the flow rate of water out of the chamber is directly proportional to the amount of water in the chamber. In fact, any system where the rate of change in a parameter is proportional to the parameter itself will have a well-defined half-life, and vice versa. Since the decay of unstable nuclei has a half-life, the rate at which the nuclei transform must therefore be proportional to the number of nuclei remaining. If nuclear decay was a collective phenomenon, this would not be so odd, because we could posit that there was some sort of interaction between nuclei that causes

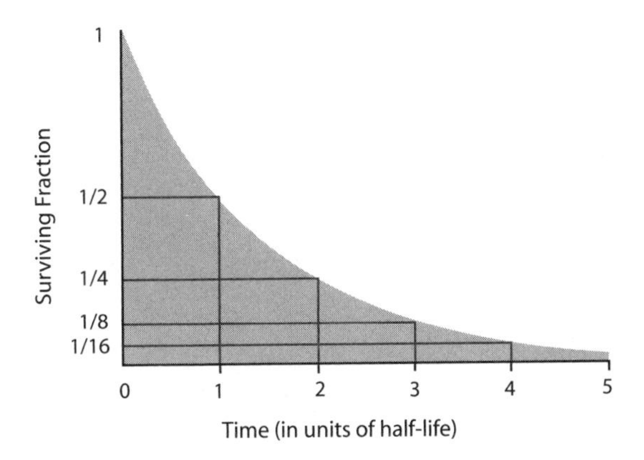

FIGURE 4.2 A decay curve, showing the fraction of undecayed material as a function of time, measured in units of half-life. After one half-life, half of the material has decayed. After another half-life, half of the remaining half has decayed, leaving one-quarter behind, and so on.

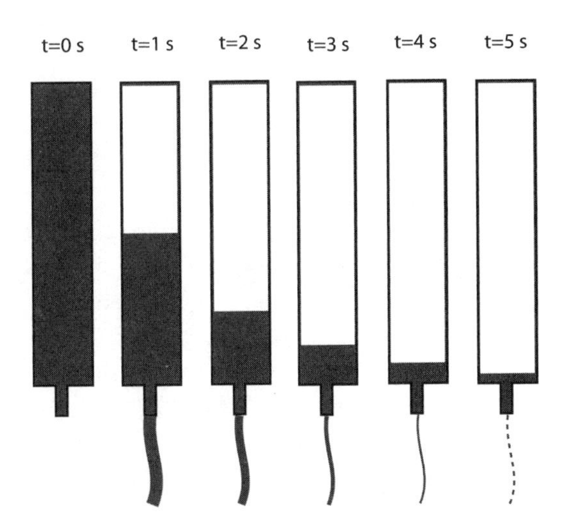

FIGURE 4.3 A system in classical physics with a half-life. This series of images shows a tank of water at a series of times. At $t = 0s$, a valve is opened at the bottom of the tank and water is pushed out by the weight of material in the tank. Initially, the flow rate is high, but as the amount of material in the tank drops, the flow slows. The images show us that the amount of water left in the tank has a half-life (compare with Figure 4.2).

them to decay faster when they are in large groups. However, this is not how nuclear decay works. Even if we ignore the experimental evidence that the relevant nuclear forces are extremely short range, the fact that the half-life of carbon-14 nuclei is always 5,700 years makes it very unlikely that interactions between nuclei could be influencing the decay rate. If such interactions did exist, then it stands to reason that we should be able to make the decay rate and the half-life change by packing the nuclei more closely together or by mixing in some other type of unstable nuclei. The half-life of carbon-14 has never been observed to vary in this way, so the decay of any given nucleus can be legitimately considered an independent event.

If each nucleus decays independently, then the parameter with a half-life must be intrinsic to each individual nucleus. However, the nucleus does not gradually change from carbon-14 to nitrogen-14, it instead transforms all at once. This decay event itself therefore cannot have a half-life. In fact, the observed half-life of a collection of nuclei arises because different nuclei decay at different times, and the probability that any given nucleus has decayed follows a curve like the one shown in Figure 4.2. Each nucleus has a 1-in-2 chance of decaying in the first 5,700 years, and if it survives that long, it has a 1-in-2 chance of making it another 5,700 years. Thus it is most accurate to say that the *probability* the transformation has not yet happened has a half-life.

If this is true, then the probability that the decay has not yet happened plays a role similar to that of the amount of water in the tank in our previous example. Just as the half-life in that situation arose because the flow rate was proportional to the water level in the chamber, here the rate of change of the probability must be proportional to the probability that the decay has not yet happened. In other words, a nucleus with more probability of being carbon-14 at a given time is more likely to transform in the next 100 years than a nucleus with a lower likelihood of being carbon-14. The probability therefore not only quantifies the chance of the decay happening, it also seems to play an active role in determining when the decay could happen. This strange situation is a great example of a quantum mechanical phenomenon.

Quantum mechanics has a reputation for being a "weird" theory, but it actually provides a perfectly reasonable, if nonintuitive, procedure for solving many physics problems. For example, say we have a particle at a well-defined position at a given initial time, and we want to know where it will be ten seconds later. The most obvious way to solve this problem is to take the particle's initial position and velocity, and then calculate the trajectory of the particle over the following ten seconds, taking into account any forces it might experience during its journey. This method is very reasonable and works quite well in situa-

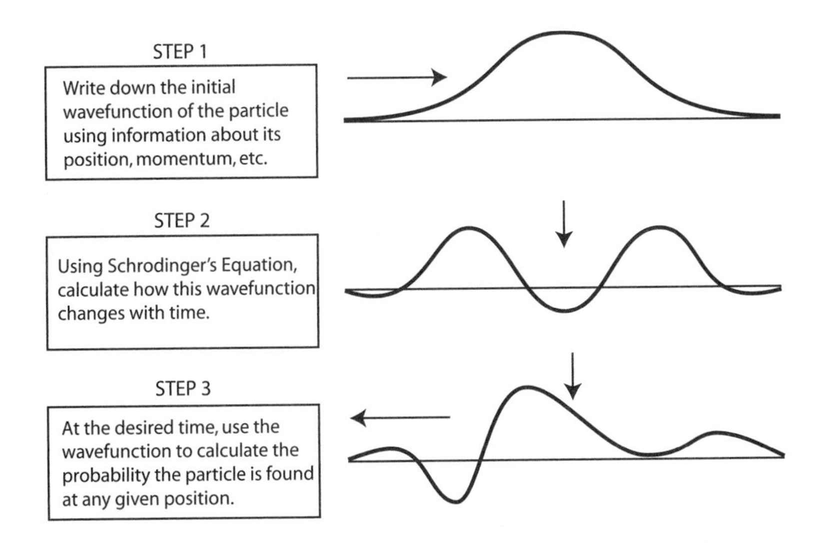

FIGURE 4.4 The usual procedure for quantum mechanical problems. A similar procedure can be used to calculate the probability that a carbon-14 nucleus has decayed after a fixed amount of time.

tions where the uncertainty in the path and final location of the particle is small. However, this procedure often fails with subatomic particles because—even if they are released from a single starting point into the same environment—they can be found at a variety of different locations after some time has passed. In this situation, we need a different procedure to determine where the particle is likely to wind up, which is provided by the theory of quantum mechanics.

The quantum mechanical approach to these sorts of problems (illustrated in Figure 4.4) begins with the calculation of the *wavefunction* based on information about the initial position and velocity of the particle. The wavefunction describes the probability that the particle is at any given point, and there are equations that tell us exactly how this wavefunction changes over time, allowing us to calculate what the wavefunction is after ten seconds. This final wavefunction then gives us the probability that the particle is found at any given position or moving in any given direction.

Quantum mechanical models yield results that match experiments, so it is a perfectly good physical theory. However, there are some obvious conceptual issues with this procedure. For instance, while we can calculate exactly how the wavefunction will evolve between its initial and final states, the final wavefunction gives only the probability that the particle is found at various locations. The process whereby this probability transforms into the specific, actual position of the particle as measured in a particular experiment is still not perfectly understood.

A thorough discussion of the subtleties of quantum mechanics is well outside the scope of this book, but even the brief introduction given here provides some insight into the physics behind nuclear decay. Just as we could make a wavefunction and compute the probability that a particle moved this way or that over the course of ten seconds, we can construct a wavefunction that gives the probability that a carbon-14 nucleus has decayed after 1 year, 100 years, 1,000 years, or 1,000,000 years. As the wavefunction evolves, the probability of the nucleus surviving as carbon-14 drops and the probability of the nucleus transforming into nitrogen-14 rises. As with the water leaking out of a chamber, the flow of the wavefunction between these two states has a half-life, which is determined entirely by the dynamics of the nucleus itself. This ebb and flow of the wavefunction means that the probability of any given nucleus surviving for a given amount of time has a well-defined half-life determined by the total number of protons and neutrons in the nucleus. A collection of carbon-14 nuclei therefore all have the same probability of decaying at different times, and the fraction of nuclei surviving has a well-defined half-life.

SECTION 4.2: GEIGER COUNTERS AND MASS SPECTROMETERS

Unstable nuclei like carbon-14 clearly have the potential to be powerful tools for measuring the passage of time. Since quantum mechanical phenomena occurring deep within individual atoms determine when the nuclei decay, we can count on the carbon-14 content of any isolated object being cut in half every 5,700 years. However, in order to exploit this simple behavior and use these nuclei as timekeepers we need to determine the carbon-14 content of an object at two points in time. Only after this is done can we calculate how much carbon-14 has decayed and the amount of time involved. In practice, researchers almost always take the present as one of these time periods, because the current carbon-14 content of objects can be measured directly.

Determining the carbon-14 content of ancient artifacts is a fairly challenging task. In a typical datable object, at most only about one out of every trillion atoms of carbon is in the form of carbon-14. These carbon-14 atoms have essentially the same chemistry as the other carbon isotopes, so they cannot be isolated using standard chemical techniques. Instead, the rare carbon-14 atoms must be identified and counted based on their unique physical properties: their larger mass and their radioactivity.

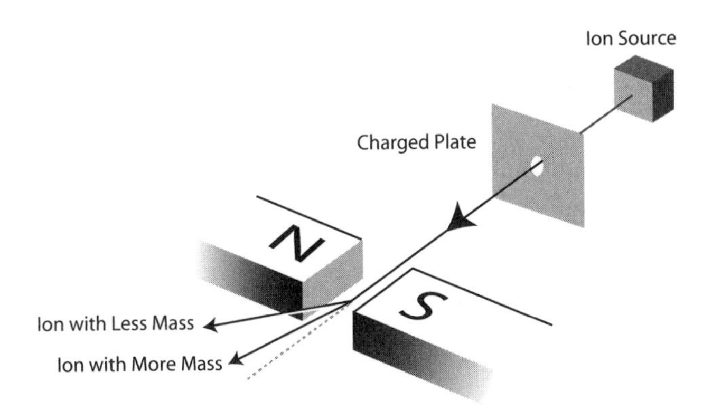

FIGURE 4.5 The basic idea of mass spectrometry. Ions generated from the sample are attracted to and accelerated by a charged plate. A hole in this plate lets the ions through to a magnetic field (here indicated schematically between north and south magnetic poles), which deflects the ions. More massive ions are deflected less, so ions with different masses can be isolated.

When Willard Libby and his colleagues first developed carbon-14 dating in the 1940s, they relied on the radioactivity of these unstable nuclei. Each time an atom of carbon-14 decays it emits an electron, which can be detected if it passes through a Geiger counter. To ensure that the electron comes from the carbon-14 in the object and not from some other radioactive element the sample must be processed and purified to isolate the carbon. Careful shielding and additional detectors are also needed in order to identify and exclude any particles coming from outside sources.

This radioactive decay method enabled Libby and others to measure the carbon-14 content of a variety of objects and was sufficient to demonstrate that carbon-14 could be used to measure age (see below). However, this method does have some serious limitations. To get a precise measure of the carbon-14 content of a given sample, we need to observe roughly 1,000 decays. Since the half-life of carbon-14 is thousands of years long, only a small fraction, 0.01%, of the carbon-14 atoms in a sample decays within a single year. This means we need ten million carbon-14 atoms in the sample to get a reasonable estimate of the carbon-14 content, and even then we still need to wait a full year. This method is therefore rather inefficient, and also requires relatively large amounts (1 gram) of carbon to work.

Nowadays, small samples of material can be dated using a technique called mass spectrometry, which uses electric and magnetic fields to sort atoms by mass (see Figure 4.5). Individual atoms are released from the sample and ionized by adding or removing electrons from each atom. These atoms have a net charge, so they are attracted toward metal plates with an opposite charge. The atoms move faster and faster as they approach the metal plates. A passage through the plates allows the atoms to go through to the other side. They then enter a magnetic field. Moving charged particles both produce magnetic fields (as in an electromagnet) and respond to external magnetic fields. The charge and velocity of the ion determine the strength of the force it feels in the magnetic field, and the mass of the atom determines how much it moves in response to this force. Atoms with different masses therefore take different trajectories through the magnetic fields, enabling the different atoms to be identified, isolated, and counted.

Standard mass spectrometers are table-top devices that are used to measure the major constituents of various materials. However, accurately measuring the extremely small fraction of carbon-14 atoms in a typical sample requires a special type of mass spectrometry, called accelerator mass spectrometry (AMS). This uses multiple stages of acceleration and ionization, as well as several magnets to cleanly separate the carbon-14 from all other possible atoms and molecules. The machines needed to do this are large beasts that fill entire buildings and exist only at specialized facilities in about half a dozen places in the United States and about two dozen other locations throughout the world.

The major advantage of AMS is that all carbon-14 atoms in a sample are counted, not just the ones that happen to decay, so this method can be used with sample sizes as small as 1 milligram. This means that artifacts can be analyzed without doing too much damage, and that even objects with small amounts of carbon (like steel tools) can potentially be dated with carbon-14.

SECTION 4.3: THE ORIGIN AND ORIGINAL LEVEL OF CARBON-14 IN LIVING THINGS

It is important to remember that the current amount of carbon-14 in an object provides us with only half the information we need to calculate its age. For example, say we find some a piece of wood from an ancient campfire. This material contains 10 micrograms of carbon-14. We can infer that this object, if it existed 5,700 years ago, had 20 micrograms of carbon-14, and if it was around 11,400 years ago, it had 40 micrograms of carbon-14. However, we cannot determine when the campfire actually happened. To find this out, we need to know the carbon-14 content of the wood when the fire occurred. Without a time machine, we can never measure this directly, but we can estimate the carbon-14 content of the wood when it was part of a living tree.

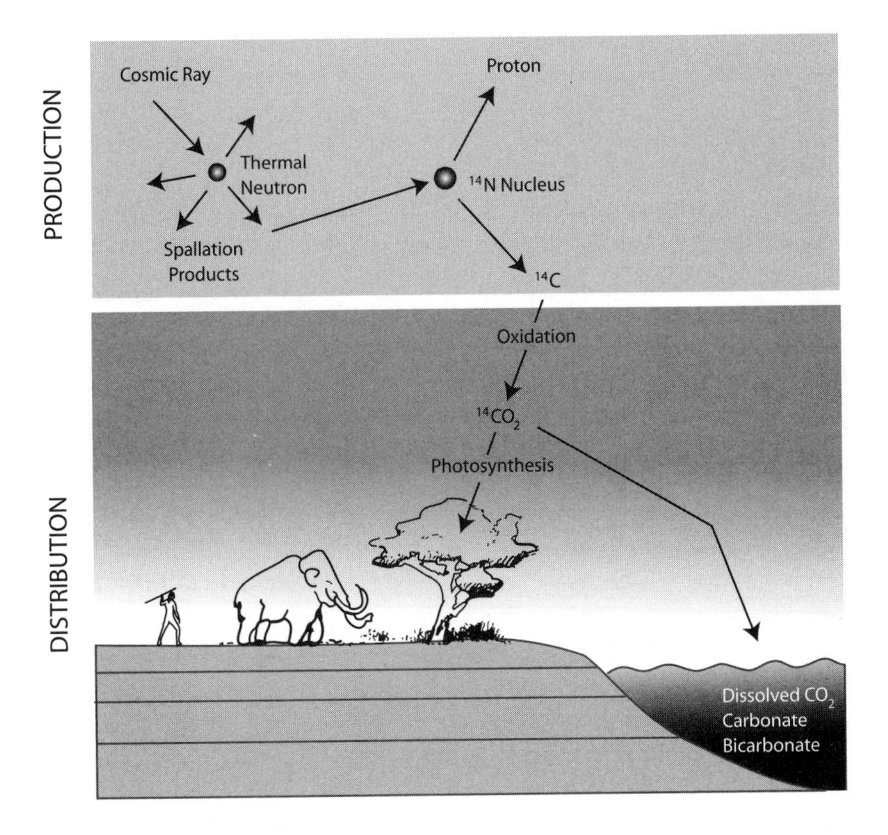

FIGURE 4.6 A summary of how carbon-14 is produced and moves through the atmosphere and various living organisms. Based on R. E. Taylor *Radiocarbon Dating* (Academic Press, 1987), Figure 1.1.

Trees, grasses, zebras, lions, people, and all manner of living things are constantly exchanging carbon atoms with each other and with the atmosphere. Plants take in carbon dioxide from the atmosphere and then use photosynthesis to convert this gas into sugars, leaves, roots, and stems. When animals eat plants or other animals, some of this material is broken back down for energy, producing the carbon dioxide that all animals exhale. Just as carbon atoms are constantly flowing between plants, animals, and the atmosphere, all of these things are also swapping carbon-14. Since the chemical properties of carbon-14 are almost identical to those of all of the other carbon atoms, this constant interchange of material distributes the carbon-14 throughout the ecosystem (Figure 4.6). All of the plants and animals alive on earth at one time therefore contain roughly the same mix of carbon isotopes that are found in the contemporary atmosphere. If the carbon-14 content of the atmosphere has remained constant over the years, then the fraction of carbon atoms in the form of carbon-14 was the same for organisms living in the past as it is for organisms living today.

There are some good reasons to expect that the carbon-14 content of the atmosphere has not changed too much with time. Until quite recently, most of the carbon-14 on earth was originally created by cosmic rays (since 1950, the carbon-14 generated by nuclear weapons testing must be taken into account, but this need not concern us here). Cosmic rays are bare atomic nuclei (i.e., atoms without electrons) that travel through the universe at extremely high speeds, many approaching the speed of light. The source of cosmic rays is still somewhat uncertain. Some come from the sun, but most come from outside our solar system. Since cosmic rays are charged particles, and there are magnetic fields in interstellar space, these particles are deflected from straight paths just like the ions in the mass spectrometer. The cosmic rays therefore follow complicated twisting paths between their source and earth, which makes it very difficult to reconstruct exactly where they came from. Some may be produced during the deaths of massive stars and others may arise from material falling into gigantic black holes. However, these particles probably come from a variety of astronomical objects. Just as we do not expect the total amount of visible light from stars to change much over the last few thousand years, we do not expect the total flux of cosmic radiation to change very quickly (but see the next chapter).

Cosmic rays produce carbon-14 when these particles collide with atoms in our atmosphere. During these collisions, the enormous kinetic energy of the cosmic ray is more than enough to break both colliding nuclei into their component parts. This energy is even sufficient to generate exotic subatomic particles. If the cosmic rays are moving fast enough, the debris from this impact can contain enough kinetic energy to produce additional violent collisions, producing a shower of nuclei and subatomic particles. This nuclear debris includes free neutrons. These neutrons rattle around in the atmosphere for a while and usually they end up stuck in the nucleus of one of the nitrogen atoms that fill the atmosphere. This happens relatively easily since the neutron does not have a charge, so it is not repelled from nuclei like a proton would be. After capturing the neutron, the resulting nucleus has seven protons and eight neutrons. This perturbed nucleus is extremely unstable and quickly spews out a proton, leaving behind an atom of carbon-14.

With cosmic rays constantly replenishing the carbon-14 in the upper atmosphere, the carbon-14 content of the air and various living creatures can remain steady over time. So long as the carbon-14 level in the atmosphere stays constant, the carbon-14 content of living organisms from any time in the past should be the same as it is today. However, once the organism dies, it can no longer obtain new carbon from the atmosphere and the carbon-14 that decays is not replaced. For example, say that a contemporary piece of wood of the same size and type as the one from our campfire has 20 micrograms of carbon-14, twice that of the ancient sample. This means that half the carbon-14 in the old wood has decayed since it stopped absorbing carbon from the atmosphere. The wood used in the campfire is therefore from a tree that died one carbon-14 half-life—or roughly 5,700 years—ago. While this does not tell us exactly when the campfire itself happened, we could presume that the death of the tree and the burning of the wood were not too far separated in time.

SECTION 4.4: EGYPTIAN ARTIFACTS AND EXPERIMENTAL VERIFICATION

Among the concepts behind carbon-14 dating described above, the most questionable assumption is that the carbon-14 content of the atmosphere remains constant in time. The decay rate of carbon-14 and the distribution of carbon-14 in the present biosphere can both be verified with modern-day measurements, but it is not as straightforward to demonstrate that there have not been radical changes in the cosmic ray flux on the atmosphere over thousands of years. The only way to prove that the carbon-14 content of the atmosphere was roughly the same in the past as it is today—and to validate the carbon-14 dating system—is to measure the carbon-14 fraction of organic material with an age that has already been well established by other means.

When Libby and his colleagues were first developing the carbon-14 dating method in the 1940s and 1950s, they measured the carbon-14 content of several objects that were dated based on historical records. Many of the objects from these studies came from Egypt, because as we saw in the previous chapter, Egypt possesses both extremely well-preserved artifacts and an unusually well-documented history. For example, a wooden boat was recovered from the tomb complex of an Egyptian king named Senusret III who ruled during Egypt's Middle Kingdom. Based on the astronomical records from this time period (described in the last chapter), we know that this boat was interred around 1840 BCE. Even the pyramids of the Old Kingdom provided useful material. Wooden beams had been used in one of the chambers of Snofru's Bent Pyramid, perhaps in an effort to keep the walls from collapsing.

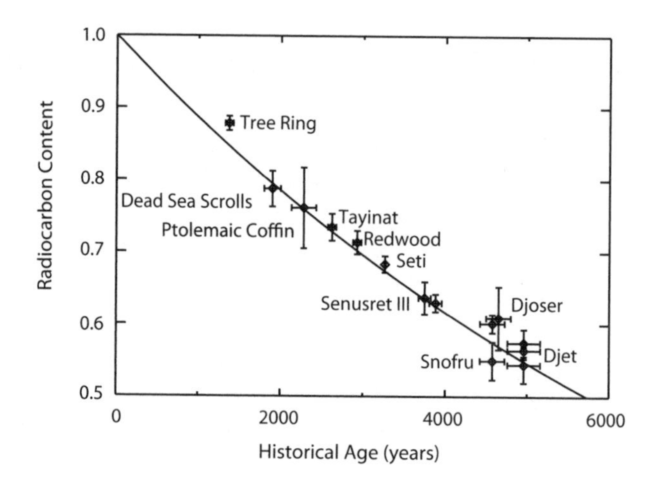

FIGURE 4.7 A "curve of knowns" that shows that older objects have less carbon-14 activity. The radiocarbon content is the ratio of carbon-14 content in an object of a given historical age to that in modern living matter. Based on Willard F. Libby, 1960 Nobel Prize Lecture in Chemistry, http://nobelprize.org/chemistry/laureates/1960/libby-lecture.html, figure 3.

In Libby's time the age of the Pyramids was uncertain by a century or two, but they were known to be nearly 5,000 years old, or almost a full carbon-14 half-life.

Figure 4.7 shows the measured carbon-14 content of these and other objects as a function of time, along with a curve that shows what we would expect to get if the carbon-14 content in the atmosphere was really a constant in time. The measured data fit this curve fairly well. The Egyptian artifacts from the age of the pyramids, which are roughly 5,000 years old, do indeed have slightly more than half the carbon-14 of contemporary living material, while more recent objects, like Senusret's boat, have higher concentrations of carbon-14. This "curve of knowns" thus indicates that the carbon-14 method can be used to measure the age of ancient artifacts and that the carbon-14 content of the atmosphere has not undergone huge changes over the last 5,000 years. However, this graph also hints at something else. Note that the older data points tend to lie a little above the line, which implies they have a little bit more carbon-14 than expected.

As carbon-14 dates became more and more precise and reached backwards to a time before historical records, it became increasingly clear that while the basic principles of carbon-14 dating were sound, the simple method outlined above needed to be refined to obtain accurate age measurements. These refinements are the subject of the next chapter, which will show that biological, archaeological, geological, and even astronomical processes all must be taken into account to realize the full potential of carbon-14 dating.

SECTION 4.5: FURTHER READING

For a general overview of carbon-14 dating, including a short history of Libby's work, see R. E. Taylor *Radiometric Dating* (Academic Press, 1987). Also see Libby's 1960 Nobel Prize lecture, available at http://nobelprize.org/chemistry/laureates/1960/libby-lecture.html

For a historical introduction to nuclear physics, try G. I. Brown *Invisible Rays* (Sutton Publishing, 2002).

For a popular introduction to the weirdness of quantum mechanics, I recommend John Gribbin *In Search of Schrödinger's Cat* (Bantam, 1984) and *Schrödinger's Kittens* (Back Bay Books, 1996), and R. P. Feynman *QED: The Strange Theory of Light and Matter* (Princeton University Press, 1988).

For those intrepid folks wanting to know how to actually solve quantum mechanics problems at college level, I recommend D. J. Griffiths *Introduction to Quantum Mechanics* (Prentice Hall, 1995).

A history of accelerator mass spectrometry can be found in H. E. Gove *From Hiroshima to the Iceman* (Institute of Physics, 1999).

For more detail about mass spectrometry than anyone would ever want, see Claudio Tuniz et al. *Accelerator Mass Spectrometry* (CRC, 1998).

CHAPTER FIVE

Calibrating Carbon-14 Dates and the History of the Air

Carbon-14 dating may be a classic example of the practical applications of nuclear physics, but extracting an age estimate from carbon-14 data is not a textbook physics problem, even if certain physics textbooks sometimes pretend otherwise. For example, during my sophomore year at Grinnell College, I was assigned the following problem in my modern physics class (radiocarbon = carbon-14):

> The relative radiocarbon activity in a piece of charcoal from the remains of an ancient campfire is 0.18 that of a contemporary specimen. How long ago did the fire occur?

I knew how I was expected to solve this sort of problem. Assuming that the carbon-14 content of the atmosphere remains roughly constant for all times, the carbon-14 content of the ancient charcoal should have originally been the same as that of the contemporary sample. Remember, carbon-14 decays with a half-life of about 5,700 years, which means that the amount of carbon-14 in the charcoal is cut in half every 5,700 years. Since 0.18 is between one-quarter and one-eighth, the charcoal must date from between two and three half-lives ago. In other words, it must be between 11,400 and 17,100 years old. Using the appropriate equations, I was able to obtain a more precise age of 14,000 years. However, after doing the calculations and writing down the result, I felt I couldn't just leave the problem like that. I had taken a few archaeology courses, and I knew that the assumptions behind this answer were not completely sound. Consequently, I scribbled down the following rant (misspellings and poor grammar included):

> This says the tree from which this sample was taken was fallen ~14,000 yrs ago. However [we] cannot judge this date as valid w/o adequate context. Does this date make sense in terms of other dates at the site? For a single date is not very reliable in these situations. Furthermore, is the sample contaminated? Could there be factors that would cause the date to be erroneus? Also the date has not been properly altered to take into account **fluctuations in radiocarbon levels** from year to year. finally, even if the above factors were taken into account, we only know when the tree was cut down (when it stopped taking up CO_2) we have no idea how long it might have sat around before being used in a fire. Therefore, unfortunately, we cannot tell when the fire occured with the limited data given.

Looking back on this rambling, semi-coherent passage today, I can understand why the professor gave me credit for solving the problem, and declined to comment on my sophomoric insight. Even so, there is a grain of truth to my paragraph: we need more information before we can accurately measure the age of the charcoal. Specifically, we need to determine the original carbon-14 content of the wood.

The physics of nuclear decay ensures that the carbon-14 atoms in any object will steadily decay in a predictable way, but there is no simple physical process that guarantees a given object had a given carbon-14 content when it was part of a living organism. Fortunately, ongoing investigations of the carbon-14 contents of different materials from different times have produced data and procedures that provide reasonably reliable estimates of the original carbon-14 content of many objects. This research has certainly improved the reliability and accuracy of carbon-14 dating; and it has also had some unexpected benefits for climatologists and astrophysicists.

One by-product of the decades-long effort to refine carbon-14 dating is a detailed record of atmospheric carbon-14 levels over the last 15,000 years. This may not seem like much, but a 1–2% increase in the carbon-14 content of the atmosphere can be the faint echo of a major change in the ocean's currents or in the sun's activity. Such changes therefore provide a unique window into the complex, interconnected processes that shape the sun's surface and the earth's climate. For example, a shift in atmospheric carbon-14 levels

13,000 years ago provides an important clue to the sequence of dramatic changes that unfolded at the end of the last Ice Age.

SECTION 5.1: FROM RAW TO CALIBRATED DATES

The standard procedure for extracting chronological information from carbon-14 can be divided into two fundamental tasks: estimate how much carbon-14 the object contained when it was part of a living creature, and determine how much carbon-14 remains in the material today. Since the current carbon-14 content of an object can be measured directly, carbon-14 dating almost always begins by obtaining the isotopic composition of a sample from the artifact. In principle, these data could be reported as a number of carbon-14 atoms in the sample, but in practice dating facilities normally compute a "raw" or "conventional" carbon-14 date for the sample. These dates are expressed as a number of years "BP," which meant "Before Present" when the present was 1950. Now that this year is several decades in the past itself, BP has been reinterpreted to mean "Before Physics."

As in the exercise I had to do in college, conventional carbon-14 dates assume the initial carbon-14 content of the ancient material was the same as a modern reference sample: specifically, a batch of oxalic acid derived from a bunch of sugar beets grown in France in 1950. As we will see below, such dates in general do not provide accurate age estimates, so today they are used only as a standardized measure of carbon-14 content. For this reason, all dating facilities follow the same—sometimes peculiar—procedures when they calculate and report a conventional carbon-14 date. For example, they take the carbon-14 half-life to be about 5,570 years—an inaccurate value used by Libby several decades ago. Such conventions may seem arbitrary, but they enforce a certain level of consistency among carbon-14 dates.

Dating facilities also make sure different materials of the same age have the same conventional carbon-14 date by accounting for a phenomenon known as mass fractionation, which allows different organisms to acquire different mixes of carbon isotopes while they are alive. The various isotopes of carbon all have the same number of protons in the nucleus and the same configuration of electrons, so they all have nearly identical chemical properties. However, different isotopes have different masses, so it requires different amounts of force to get them moving. The processes that transport carbon atoms from one location to another can therefore move some isotopes more efficiently than others, producing either enhancements or depletions of heavier isotopes in different locations. This means that some creatures can accumulate more carbon-14 in their bodies than others. For example, certain plants like corn use a slightly different photosynthetic process from other plants to absorb carbon from the atmosphere, which causes living corn plants to have a slightly (2–3%) higher carbon-14 fraction than tree leaves or sugar beets growing at the same time. If we neglect this phenomenon, carbon-14 dating will underestimate the age of materials derived from corn.

Scientists account for the effects of mass fractionation by measuring the relative amounts of stable carbon isotopes in the sample. Most stable carbon, also known as carbon-12, has 6 protons and 6 neutrons. However, roughly one percent of the stable carbon atoms are in the form of carbon-13, which has 7 neutrons. These two isotopes have slightly different masses, so any process that sorts the carbon atoms by mass will not only alter the carbon-14 fraction of the material but also change the mix of carbon-12 and carbon-13. The relative amounts of carbon-12 and carbon-13 can then be used to correct for any mass fractionation affects in the carbon-14 content.

For example, say the standard French sugar beets had the following mix of isotopes: 99 parts carbon-12, 1 part carbon-13, and 0.000,000,000, 1 parts carbon-14. Imagine we want to determine the age of a piece of an ancient basket that contains 98.9 parts carbon-12, 1.1 parts carbon-13, and 0.000,000,000,06 parts carbon-14. The carbon-14 fraction of the basket is about six-tenths—a little more than one-half—the carbon-14 fraction of the sugar beets. If we neglected mass fractionation, we would estimate that the basket is a little less than a half-life, or about 5,500 years old. However, since the sugar beets and the basket have different ratios of carbon-12 and carbon-13, we can be sure that mass fractionation has occurred. Since the basket has 10% more of the heavier stable isotope, it also should have had more carbon-14 than the sugar beets originally. The mass difference between carbon-12 and carbon-13 is half the mass difference between carbon-12 and carbon-14, so the mass fractionation should be twice as strong for carbon-14. The initial carbon-14 fraction of the basket should therefore be 20% higher than the carbon-14 fraction of sugar beets living at the same time. Assuming sugar beets then and now had the same carbon-14 content, the original carbon-14 fraction of the basket was 0.000,000,000, 12%, exactly twice its current carbon-14 content. This means we must revise our estimate of the age of the basket to one full half-life, or 5,700 years.

While including corrections for mass fractionation ensures that different organisms living in the same area at the same time will have similar carbon-14 dates, we still need to determine what a conventional carbon-14 date corresponds to in real time. To accomplish this, we need some additional infor-

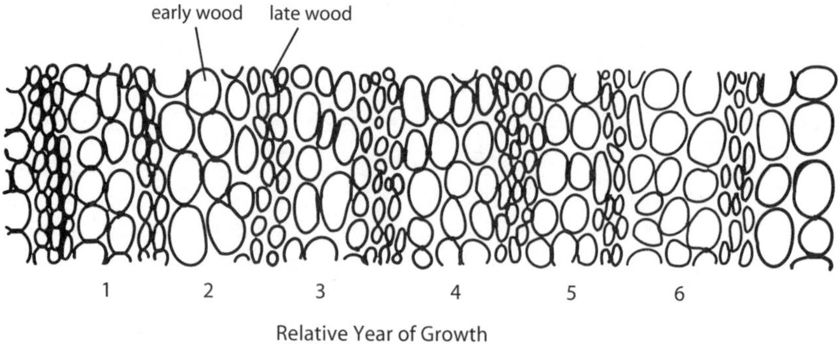

FIGURE 5.1 Schematic tree rings. Each year, the wood growing under the bark of a tree starts out with an open structure to allow maximum water flow up from the roots to the growing leaves. Later in the year, the material becomes denser. This pattern repeats annually in temperate environments.

mation about the past carbon-14 content of the air. As we discussed in the last chapter, all living things ultimately obtain their carbon-14 from the atmosphere. If atmospheric carbon-14 levels have remained constant over the last few thousand years, then an organism living any time in the past would have had the same carbon-14 content as a similar organism living today, and conventional carbon-14 dates would yield proper estimates of age (assuming we use the correct half-life). Libby's data from ancient Egyptian artifacts show that the carbon-14 content of the atmosphere has been fairly stable over the last few thousand years. Still, these carbon-14 dates are somewhat less than their historical age, which suggests that there may have been changes in the amount of carbon-14 in the atmosphere. If the carbon-14 content of the atmosphere fluctuates, then the age estimates based on conventional radiocarbon dates will be off. We therefore need to determine what the isotopic composition of the air was like in the past before we can use radiocarbon dating techniques to explore the history of the people or animals which breathed that air.

Fortunately, the history of atmospheric carbon-14 levels is recorded in tree rings, the familiar pattern of dark and light bands that can be seen in any cut piece of wood. This material represents a part of the tree's vascular system that carries water up from the roots to the leaves (see Figure 5.1). Only the outermost layers of this tissue actively carry fluids, and new layers are constantly being added to the tree underneath the bark of the trunk. In spring, as new leaves are growing, the demand for water is high and the wood has an open structure. As the season progresses, the need for water declines and the tissue becomes denser with fewer open spaces for water to flow through. This continues until winter, when the tree can become dormant until the next spring and the cycle starts all over again. Each ring—consisting of a light band and dark band—therefore corresponds to precisely one year of growth in a temperate environment, and we can determine how old a tree is by counting its rings.

If we study these rings in more detail, we will find that they can provide more information than just the tree's age. Tree rings have different thicknesses depending on how much the tree grew from year to year. Since the quality of the growing season depends on local weather conditions, the same pattern of ring thicknesses will occur in many of the trees in the same region. By matching the patterns in trees of different ages, scientists can generate a continuous record of tree rings, assigning a specific year to each and every ring (see Figure 5.2). For example, say we chop down a tree today and look at its rings. Since

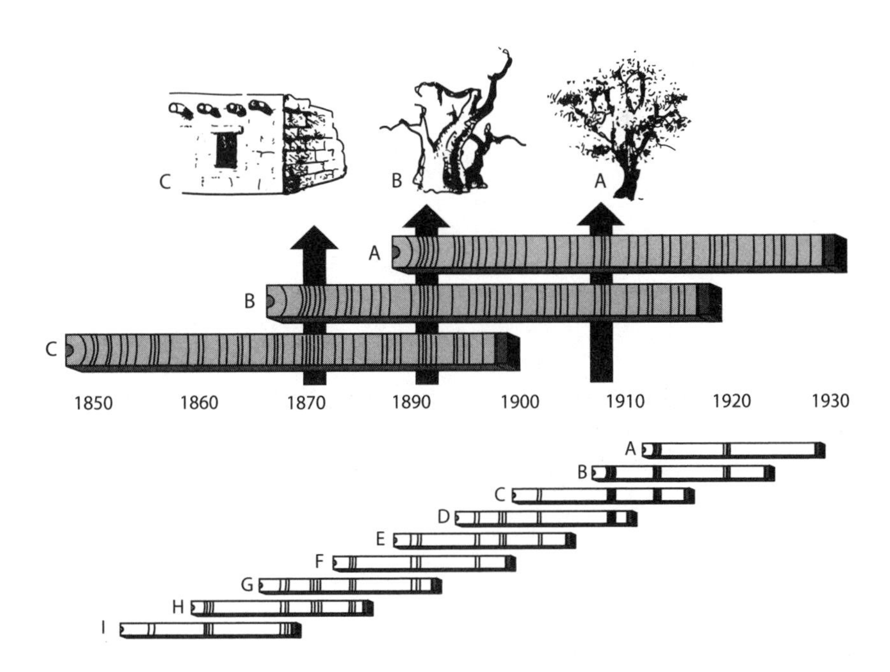

FIGURE 5.2 The basic idea of dendrochronology (based on a figure from R. E. Taylor and M. J. Aitken *Chronometric Dating in Archaeology* [Plenum Press, 1997]). Pieces of wood from (A) a living tree, (B) a dead tree, and (C) a beam from a house that grew in the same area reveal patterns in the thickness and spacing of tree rings that indicate that these particular rings were formed at the same time. By matching such patterns in many pieces of wood, the record of tree rings can be extended far into the past.

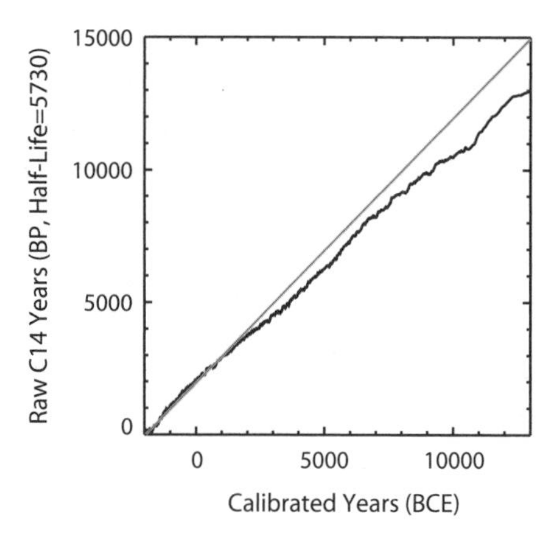

FIGURE 5.3 The calibration curve for carbon-14 dates, as of 2004. This curve gives the carbon-14 age as a function of real years, measured using dendrochronological and other data sets. For simplicity, the "raw" dates are given using the proper half-life instead of the conventional value. If carbon-14 dating were perfectly accurate, then the data would fall on the straight line. In fact, before 2000 BCE, conventional carbon-14 dates significantly underestimate the actual age of the specimens. Data obtained from www.radiocarbon.org/IntCal04.htm.

we know the outermost ring corresponds to this year, we can figure out exactly when each and every ring in that log was made. Now if we find a second log from the same area, we can look to see if there is a pattern of ring thicknesses that matches the rings in the original log. If there is, then we know when all the rings in the second log formed, including those that were made before the first tree even existed. Repeating this procedure over and over again yields extremely long sequences of rings, stretching back thousands of years.

Using logs from Germany, Ireland, and the west coast of the United States, researchers have managed to construct continuous records of tree rings all the way back to about 10000 BCE. Other cyclic natural phenomena, such as layered deposits produced by the seasonal fluctuations in river runoff, can provide similar records extending back even further into the past. By measuring the carbon-14 content of this material, scientists can relate conventional radiocarbon dates to actual ages. An international consortium regularly reevaluates these data and combines them to produce a standard estimate of the actual age corresponding to any particular radiocarbon date. These results are often displayed as a graph like that shown in Figure 5.3, which shows the conventional carbon-14 dates as a function of real age. Such a curve both reveals inaccuracies in the conventional carbon-14 dates, and provides the information needed to transform a carbon-14 measurement into a proper age estimate.

Consider a tree ring that has half the carbon-14 fraction of a contemporary sample. The raw carbon-14 age of this ring would indicate that it is 5,700 years old. The data in Figure 5.3 shows that a tree ring with a carbon-14 age of 5,700 years comes from about 4500 BCE, and is therefore actually about 6,500 years old.

Since all living things ultimately obtain carbon-14 from the atmosphere, the remains of any organism that lived 6,500 years ago should have the same conventional carbon-14 date as the above tree ring (after accounting for mass fractionation, etc.). We can therefore use this calibration curve to convert *any* raw carbon-14 age estimate into real years. In principle, all we need to do is find the carbon-14 date on the vertical axis of the plot and draw a horizontal line across the graph(see Figure 5.4). The place where the calibration curve intersects that line gives the real age in years of the sample. This age estimate

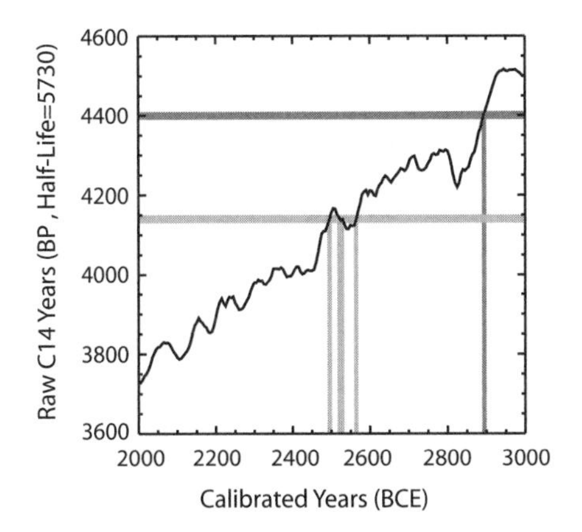

FIGURE 5.4 Using the tree ring dates to calibrate carbon-14 data. Here we show a small portion of the curve illustrated in Figure 5.3. For simplicity, the raw dates are given using the proper half-life instead of the conventional value. Say we get a carbon-14 date of 4400 BP; then we can draw a horizontal (dark gray) line across the plot and find where that line intersects the curve. Then we go down and find the age of the sample in real years, which in this case is 2900 BCE. The wiggles in the calibration curve, however, can complicate this estimate. Say that the raw carbon-14 date was 4140 BP (illustrated with the light gray line); then there are three possible values for the age of the sample, so even if the uncertainty in the carbon-14 date is only 20 years, the uncertainty in the actual age of the sample is over 100 years.

is called a calibrated carbon-14 date, and is often given as a number of years "cal BP" or as a date in the familiar AD/BC (or CE/BCE) system.[1]

It is important to note that the various wiggles in the curve sometimes mean that multiple calendar dates are consistent with the same conventional carbon-14 date. In these situations, the multiple possibilities for the age of the sample tend to amplify the uncertainty in an age measurement. For example, in Figure 5.4, a 20-year uncertainty in the carbon-14 date yields a true age estimate with an uncertainty of about 100 years. This can greatly complicate efforts to precisely date material from certain time periods.

SECTION 5.2: THE CALIBRATION CURVE AS A HISTORICAL RECORD

The calibration curve shown in Figure 5.3 is far more than a useful tool for archaeologists. It also preserves a detailed record of the carbon-14 content of the atmosphere. For example, the 6,500-year-old tree rings have conventional carbon-14 dates of about 5,700 years, which means they contain approximately half as much carbon-14 as a contemporary sample. Since these tree rings are well over one half-life old, they must have originally contained *more* than twice as much carbon-14 than they do now. These rings therefore must have contained more carbon-14 when they formed than similar tree rings growing today. Like all other living things, trees obtain carbon-14 from the atmosphere, so the atmosphere also must have contained significantly more carbon-14 6,500 years ago than it does today (see Figure 5.5). This could indicate that carbon-14 was being produced at a higher rate in the past because more cosmic rays were colliding with the upper atmosphere. Alternatively, perhaps carbon-14 is being removed from the air more quickly now than it was 6,500 years ago due to some shift in the global carbon cycle. In fact, recent research demonstrates that atmospheric carbon-14 levels have been influenced by events both here on earth and in outer space. The calibration curve can therefore provide important information about how our astrophysical and climatic environments have changed over the last 15,000 years.

While the cosmic rays responsible for producing carbon-14 come from deep space, thus far we have no evidence that events happening outside the solar system have influenced carbon-14 levels in the atmosphere. Instead, many of the observed fluctuations in the air's carbon-14 content can be attributed to changes in the magnetic fields surrounding the earth. Remember that cosmic

1. However, note that different scientists sometimes use other conventions.

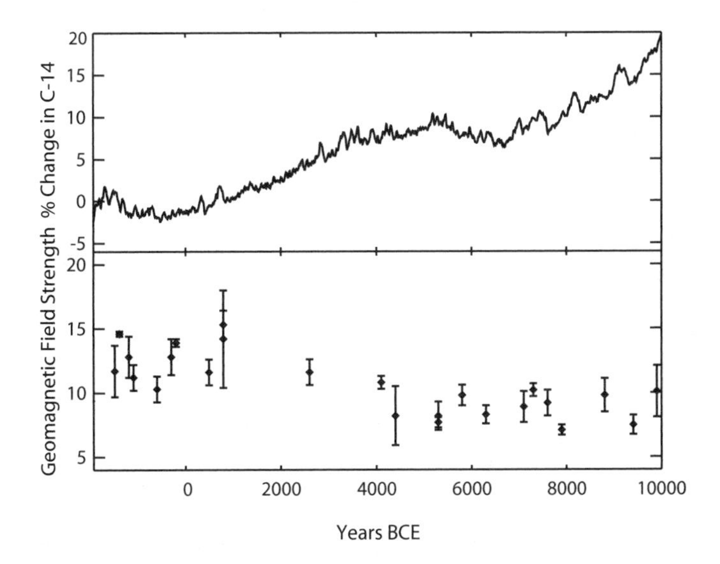

FIGURE 5.5 Top: the variations in the original carbon-14 content of the atmosphere over time inferred from the calibration curve. Note that in the past, the atmosphere often contained more carbon-14 than it does today. Bottom: the variations in the earth's magnetic field as a function of time, measured using lava flows in Hawaii (data in units of 10^{22} Am^2 from Carlos Laj et al., "Geomagnetic Intensity and Inclination Variations at Hawaii for the Past 98 kyr from Core SOH-4 (Big Island): A New Study and Comparison with Existing Contemporary Data" *Earth and Planetary Sciences* 200 (2002): 177-190). Note that before about 2000 BCE, the strength in the earth's magnetic field was lower, which may help explain why the carbon-14 content in the atmosphere was higher at that time.

rays are charged particles, so they can be deflected by magnetic fields. Shifts in the state or strength of these fields will therefore influence the number of cosmic rays from interstellar space that manage to make it to earth.

For example, consider the earth's own magnetic field, which has a reasonably well-documented history preserved in volcanic rocks. These ancient lava flows contain small grains of material that became magnetized when they solidified, and the strength of this magnetization is proportional to the strength of the geomagnetic field at that time. In places like Hawaii, where there are many lava flows with a range of ages,[2] researchers have been able to document how earth's magnetic field strength has changed over the years. Some of these

2. Geologists are able to determine the ages of these rocks using techniques described in chapter 7.

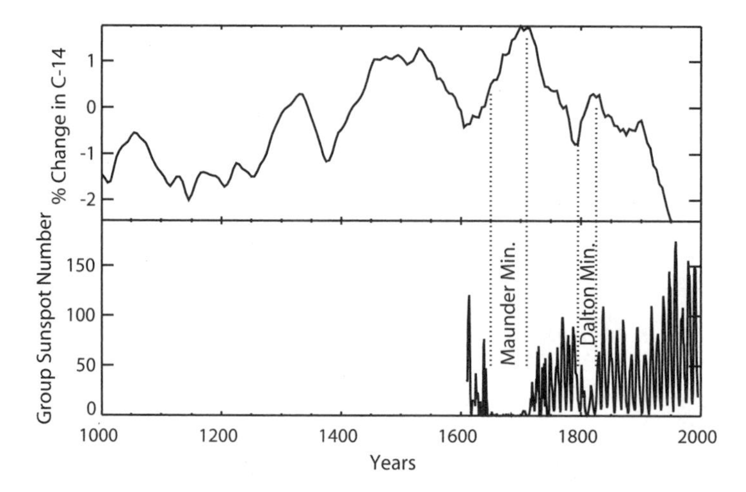

FIGURE 5.6 Top: the fluctuation in the carbon-14 content of the atmosphere for the last 1,000 years. Bottom: a standardized measure of the number of sunspots since 1600. The sunspot number rises and falls roughly every 11 years, a period known as the solar cycle. There are also two times, one around 1820 and one around 1690, when few sunspots were seen, known as the Dalton minimum and the Maunder minimum respectively. These dates coincide with increases in the atmospheric carbon-14 content that likely reflect changes in the state of the solar magnetic field.

data are illustrated in Figure 5.5, and they show that the magnetic field got significantly stronger around 2000 BCE. At about the same time, the carbon-14 content of the atmosphere drops, as we would expect if the stronger magnetic field reduced the number of cosmic rays reaching our atmosphere.

Unlike volcanic rocks, carbon-14 provides only an indirect measurement of past geomagnetism. However, atmospheric carbon-14 not only is sensitive to earth's magnetic field, it also appears to be influenced by the sun. If we take a close look at Figure 5.6, which depicts the calibration data over the last millennium, we can see there are peaks in the carbon-14 content of the atmosphere around the years 1050, 1350, 1500, 1700, and 1820. Interestingly, the last two of these periods appear to follow times of unusual solar activity.

Beginning about 1600, people have recorded the presence of the small blemishes on the surface of the sun know as sunspots. From these observations, we know that the number of sunspots rises and then falls about every eleven years. However, the number of sunspots witnessed in each cycle varies. In particular, during the latter part of the 1600s few sunspots were seen—

this period is now known as the Maunder minimum. A similar, though less extreme, reduction in sunspot number occurred around 1820 and is called the Dalton minimum. Intriguingly, both these minima just precede two of the periods when carbon-14 levels in our atmosphere peaked.

Data gathered over the last century confirm that this relationship between sunspot activity and atmospheric carbon-14 is probably not just a coincidence. More sophisticated observations of the sun have revealed that sunspots correspond to areas of intense magnetic fields. Furthermore, we now know that each time the number of sunspots rises and falls, the entire magnetic field of the sun undergoes a major reorganization. It is therefore likely that during the Maunder and Dalton minima, the magnetic field of the sun was in a different state than it typically is today, a state that apparently allowed more cosmic rays from deep space to reach our atmosphere, driving up carbon-14 production. A few decades later, when the sunspots reappear, the magnetic field of the sun presumably returned to its previous state, reducing the cosmic ray flux and eventually slowing the formation of carbon-14.

At the present moment, the processes that cause the sun's magnetic field to realign itself every eleven years are not perfectly understood. It is also not clear what happened to the sun during the Maunder and Dalton minima. This is because the outer layers of the sun are a churning mass of superheated plasma, in which nuclei and electrons can travel separately from one another. These charged particles are deflected by preexisting magnetic fields, but at the same time the motions of these charged particles can generate magnetic fields of their own. The motions of the plasma and the state of the magnetic field are therefore intertwined in a very complex and dynamic way that is difficult to model and predict. While ever more detailed observations of the sun and more sophisticated computer models are beginning to unravel the dynamics of the sun's outer layers, the carbon-14 data can also play an important role in this area.

Unlike the sunspot records, which document only two pronounced minima, the carbon-14 data cover a much larger timescale and document many more peaks in carbon-14 content, most of which probably represent additional periods of altered solar activity. Indeed, some scientists have recently used the carbon-14 calibration data to produce a record of solar activity for the last ten thousand years. According to this analysis, several of the relatively narrow blips in the atmospheric carbon-14 levels, including one from around 800 BCE and several more between 3000 and 4000 BCE (see Figure 5.5) reflect periods of unusual solar activity like those associated with reduced sunspot numbers. At present, the implications of these earlier events are obscure, but hopefully further analysis of these data will help elucidate how long periods like the Maunder and Dalton minima typically last and how often they occur, which will likely shed new light on how our star functions.

At the same time as astrophysicists are using the carbon-14 data to study the sun, other scientists are using them to gain insights into changes in earth's climate. Remember that the total amount of carbon-14 in the atmosphere at any given time depends not only on how fast carbon-14 is generated by cosmic rays, but also on how quickly carbon-14 is taken out of the atmosphere. Carbon-14 leaves the atmosphere when living things die and, more importantly, when it passes into the oceans. Changes in life on earth or in the condition of the oceans can therefore have an impact on the carbon-14 content of the atmosphere. Of course, before we can use the tree ring data to understand climatic changes, we need some method of distinguishing changes in carbon-14 levels due to climate shifts from those due to fluctuations in solar activity or the geomagnetic field. Fortunately, the data from glaciers makes this possible, at least in certain situations.

Glaciers form as layer upon layer of snow falls over cold regions like Greenland or Antarctica. Deeper layers of the glacier therefore correspond to older layers of snow, and drilling down through the glacier provides a continuous record of ice formation that can extend back tens of thousands of years. Like the carbon-14 in tree rings, the chemical composition of the ice in different layers of the glacier is sensitive to the cosmic ray flux on earth and the prevailing climate. The mix of oxygen isotopes and the amount of beryllium-10 contained in the ice are particularly informative.

Beryllium-10 is an unstable isotope that—like carbon-14—is produced in the upper atmosphere by cosmic rays. However, it has different chemical properties that prevent it from participating in the same biochemical or climatic phenomena that impact carbon-14. Much as the carbon-14 content of the atmosphere can be extracted from tree rings, the concentration of beryllium-10 in the atmosphere can be computed from ice cores. As we can see in Figure 5.7, the two curves show similar features: many of the peaks in the carbon-14 data—such as those from around 5,500 years ago—occur at the same time as peaks in the beryllium-10 record. Since both of these atoms are produced by cosmic rays, this confirms that many of these peaks are due to changes in the cosmic ray flux on the upper atmosphere, most likely because of variations in solar activity like those found in the Maunder minimum.

However, there are also some noticeable differences between these two curves, and the data between 12,000 and 13,000 years ago is particularly interesting. Here are peaks in both the carbon-14 data and the beryllium-10 data,

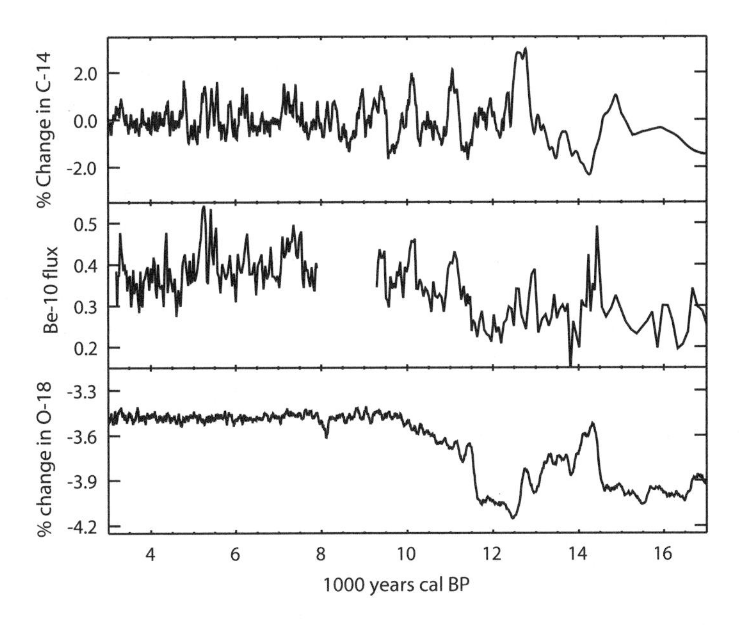

FIGURE 5.7 Changes in the carbon-14 content of the atmosphere (top), compared with variations in the beryllium-10 content (middle) and the oxygen-18 content (bottom) derived from ice cores. To facilitate comparisons, long-term drifts have been removed from the carbon-14 data. Note that many of the peaks in the carbon-14 data are mirrored in the beryllium-10 data, which confirms that both are the result of changes in cosmic ray intensity. However, the peak in the carbon-14 data around 12,700 years ago differs from the peaks in the beryllium-10 data. This peak may be related to the sudden climate shift indicated by the oxygen isotope ratios. Higher amounts of oxygen-18 indicate warmer conditions, and there is a pronounced shift to colder temperatures around this time. The beryllium-10 fluxes are given in units of 10^6 atoms/cm^2/year and come from the GISP-2 ice core (shifted in time slightly to align with new C-14 data), while the oxygen isotope data derive from the GRIP ice core. These data are published in *Journal of Geophysical Research* 102, no. C12 (1997) and were provided by the National Snow and Ice Data Center, University of Colorado at Boulder, and the WDC-A for Paleoclimatology, National Geophysical Data Center, Boulder, Colorado, http://www.ncdc.noaa.gov/paleo/icecore/greenland/summit/index.html.

but the most prominent peak in the carbon-14 data is larger and does not line up with any of the beryllium-10 peaks precisely. This peak therefore does not appear to be associated with a change in solar activity. A clue to the origin of this feature can be found in another part of the ice core data, the oxygen isotope ratios.

Water molecules are composed of two atoms of hydrogen and one atom of oxygen, so ice is rich in oxygen. Like carbon, oxygen appears in several different isotopes which have the same chemistry but slightly different masses. These isotopes also undergo mass fractionation in various situations. For example, if water containing the isotopes oxygen-18 and oxygen-16 evaporates from the surface of a lake or ocean, oxygen-16 will move into the air more quickly because it takes a little less energy for it to escape from the liquid. This difference hardly matters it is hot, because there is enough thermal energy to release water containing either oxygen-18 or oxygen-16. When the temperature drops, however, the difference between oxygen-16 and oxygen-18 becomes more important because the thermal motions of the particles are reduced. For this and other reasons, the mix of oxygen-18 and oxygen-16 in the atmosphere changes with temperature. This ratio can therefore serve a sort of thermometer for studying past climates.

The bottom panel of Figure 5.7 shows the change in the oxygen-18 concentration obtained from the same glacier in Greenland as the beryllium-10 data. Over the last 10,000 years, the oxygen isotope ratio has been fairly steady, indicating that the climate has been fairly constant. However, between 10,000 and 15,000 years ago there are several dramatic shifts in the oxygen isotope ratio. These shifts correspond to a series of climatic changes that occurred at the end of the last Ice Age. Back beyond 15,000 years ago there was less oxygen-18 in the snow falling onto the glacier, which reflects the significantly colder prevailing temperatures during the Ice Age itself. Around 14,500 years ago, the oxygen-18 ratio climbs, marking a time of pronounced warming. This warm patch lasted about 1,000 years before the onset of the so-called Younger Dryas, a 1,000-year-long period of cold temperatures in the Northern hemisphere that occurred just before the climate warmed up to its present state.[3]

The large peak in the carbon-14 content of the atmosphere around 12,700 years ago occurs at roughly the same time as the pronounced cooling that marks the beginning of the Younger Dryas. This coincidence strongly suggests that some climatic phenomenon associated with the temperature decrease provoked a significant change in atmospheric carbon-14 levels. There are only a limited number of events that could affect the atmosphere's composition so strongly, and in this case the most likely culprit is a massive shift in ocean currents.

3. The Older Dryas is a shorter period of cooling that occurred around 14,000 years ago.

Today, powerful currents like the Gulf Stream carry water between deep and shallow layers of the oceans. This mixing of the oceans enables carbon-14 to become distributed throughout the ocean waters much more efficiently than would be the case if had to diffuse down the bottom of the ocean on its own. It is possible that when the ice sheets began to melt at the end of the Ice Age, they dumped so much cold fresh water into the oceans that they disrupted these currents. This could have slowed the uptake of carbon-14 into the ocean, resulting in an excess of it in the atmosphere. Support for this model was recently found in the carbon-14 content of deep-water corals, which can trace the carbon-14 content of the deep ocean like tree rings do the carbon-14 content of the atmosphere. These data show that the deep oceans contained less carbon-14 during the Younger Dryas than they did during earlier and later warm time periods, which is what we would expect if there was less mixing between the ocean and the atmosphere at this time.

Dramatic changes in the ocean's circulation would have serious effects on the global climate, and was likely responsible for the cooling that occurred during the Younger Dryas. Earth's climate is complex, however, and there are many unresolved questions about the exact sequence of events that unfolded at the end of the last Ice Age. For example, episodes of glacier melting before and after the Younger Dryas led to significant increases in sea levels but apparently did not disrupt ocean currents or produce a hemisphere-scale temperature drop. What then was special about the conditions 13,000 years ago that led to such radical climatic shifts? Climatologists are currently working to answer such questions using more and more detailed studies of the climatic data, including information from carbon-14. For example, sufficiently precise carbon-14 dates associated with geological features like run-off channels from various glaciers may let researchers determine whether water released from a particular region triggered the onset of the Younger Dryas.

While climatologists are using the carbon-14 calibration data to figure out what the climate was like at the end of the last Ice Age, archaeologists rely on carbon-14 dating itself to study how people lived during this period. For example, humans first arrived in North and South America at least 13,000 years ago, and carbon-14 dates supply critical data about this early stage of the history of the New World. Yet, as we will see in the next chapter, there is still much controversy and occasionally fierce debates about the reliability of certain carbon-14 dates and what they imply about the people living in the Americas over 10,000 years ago.

SECTION 5.3: FURTHER READING

For a general discussion of different methods of measuring age in archaeology, including dendrochronology and carbon-14, try R. E. Taylor and M. J. Aitken *Chronometric Dating in Archaeology* (Plenum Press, 1997).

For the latest on efforts to calibrate carbon-14 dates, check out the journal *Radiocarbon* and the website www.radiocarbon.org. The most recent calibration data are published in vol. 46, no. 3 (2004) and are available on the website.

An entertaining discussion of the general chaos that carbon-14 dating caused in old-world archaeology can be found in Colin Renfrew *Before Civilization* (Knopf, 1973).

For more details about solar activity and its variations, an approachable general book is D. G. Wentzel *The Restless Sun* (Smithsonian, 1989). The website www.spaceweather.com also contains useful data about current solar activity and a variety of links. A more detailed discussion appropriate for those with some physics background can be found in Peter Wilson *Solar and Stellar Activity Cycles* (Cambridge, 1994).

A recent effort to use carbon-14 to infer solar activity is found in S. K. Solanki et al. "Unusual Activity of the Sun during Recent Decades Compared to the Previous 11,000 Years" *Nature* 431 (2004): 1084–1087.

For current efforts to model and predict solar activity, see Stuart Clark "The Dark Side of the Sun" *Science* 441 (2006): 402 and the references therein.

For more information about the Ice Age, try Richard Foster Flint *Glacial and Quaternary Geology* (John Wiley and Sons, 1971) and Jurgen Ehlers *Quaternary and Glacial Geology* (John Wiley and Sons, 1996).

For the ice core data, see the special issue of the *Journal of Geophysical Research*, vol. 102, no. C12 (1997) on the Greenland Ice Sheet Project 2 (GISP-2) and Greenland Ice Core Project (GRIP).

For the possible link between the carbon-14 data and the climate changes at the end of the last Ice Age, see K. Hughen et al. "Synchronous Radiocarbon and Climate Shifts during the Last Glaciation" *Science* 290 (2000): 1951, Raimund Muscheler et al. "Changes in Deep-Water Formation during the Younger Dryas Event Inferred from ^{10}Be and ^{14}C Records" *Nature* 408 (2000): 567–570, Laura F. Robinson et al. "Radiocarbon Variability in the Western North Atlantic during the Last Deglaciation" *Science* 310 (2005): 1469–1473, and Stein Bondevik et al. "Changes in the North Atlantic Radiocarbon Reservoir Ages during the Allerod and Younger Dryas" *Science* 312 (2006): 1514–1517.

For more information about the possible origins of the Younger Dryas, see W. Broecker "Was the Younger Dryas Triggered by a Flood?" *Science* 312 (2006) 1146–1148 and the references cited.

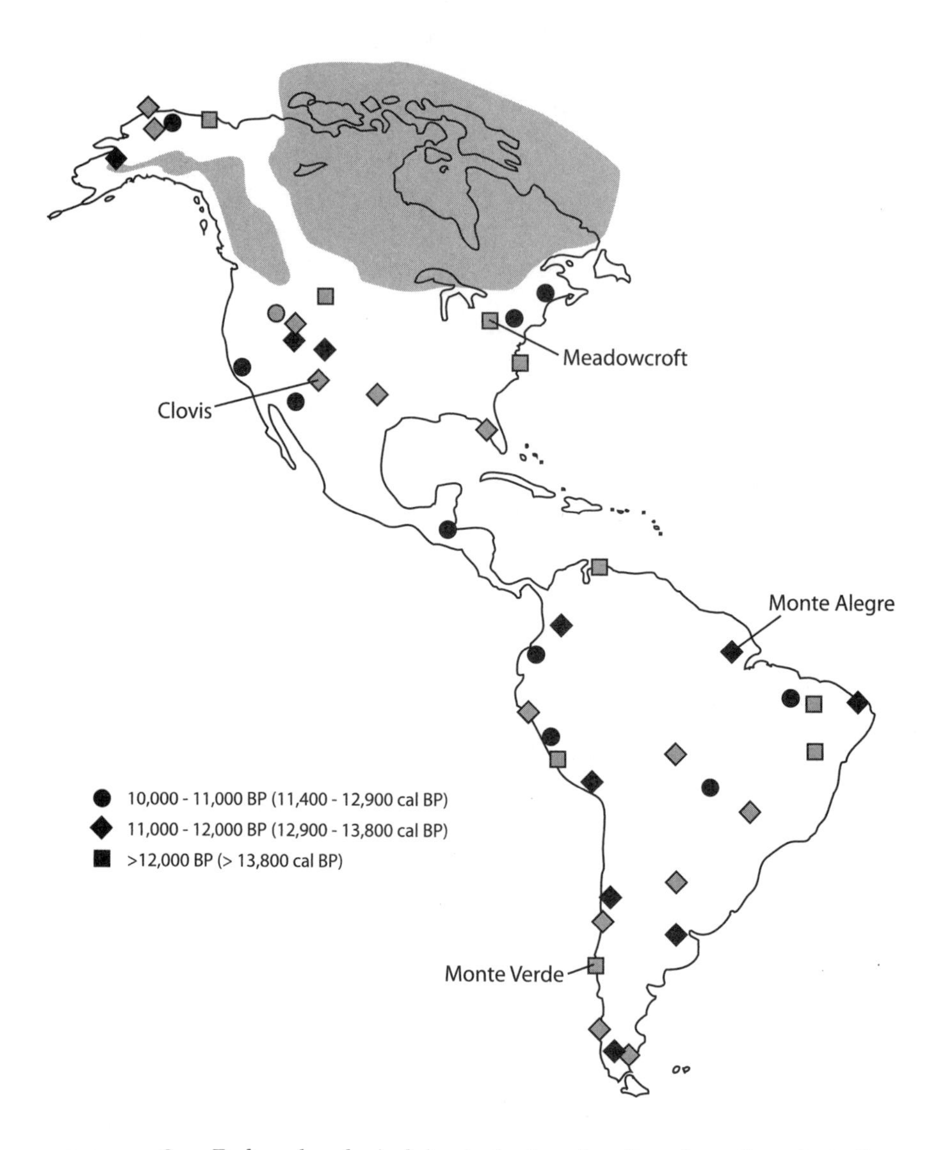

FIGURE 6.1 Early archaeological sites in the Americas (based on a figure in A. C. Roosevelt, J. Douglas, and L. Brown "The Migrations and Adaptations of the First Americans: Clovis and Pre-Clovis Viewed from South America" in *The First Americans* edited by N. G. Jablonski (University of California Press, 2002)). Gray symbols indicate uncertain or contentious dates. The shaded regions in the north show the extent of the glaciers at 12000 BP. Sites mentioned in the text are labeled.

CHAPTER SIX

Carbon-14 and the Peopling of the New World

For thousands of years, the continents now called North and South America have been host to a large array of different cultures, from small bands of hunters and gatherers to vast empires incorporating millions of people. Most of these groups, unlike the Classic Mayans or the ancient Egyptians, did not leave behind many written records. Fortunately, the tools, artifacts, and other physical remains that have been preserved provide us with insights into many aspects of these people's lives and experiences. The remains of ancient houses can tell us how many people lived under a single roof, pottery vessels contain clues about what people ate, and chips of stone can document long-distance trade routes. It is even possible to trace changes in how and where people lived over time, thanks in no small part to the chronological information contained in carbon-14 dates. As archaeologists probe further back in time, the material available to study typically becomes sparser and more fragmentary, making it more difficult to unravel how people lived in the very distant past (Fig. 6.1). It should therefore not be surprising that one of the most contentious issues in American archaeology today involves the earliest inhabitants of the New World.

There is solid evidence of human activity in the Americas as far back as the end of the last Ice Age 13,000 years ago. The debates and controversies involve several archaeological sites that suggest people may have been living throughout the New World thousands of years earlier than this, during the height of the Ice Age itself. The climatic conditions and environmental barriers along various routes to the New World at this time were very different than they were even 13,000 year ago. Ice sheets were more extensive, sea levels were lower, and so on. Therefore, if we wish to understand how and when humans arrived in North and South America, we need to determine whether the carbon-14 dates associated with these early sites are accurate and reliable.

SECTION 6.1: CLOVIS POINTS AND THE PEOPLE WHO MADE THEM

For many years, discussions of how people first reached the New World have revolved around a group of artifacts called Clovis points (shown in Figure 6.2), named after a town in New Mexico. These objects resemble large arrowheads, but in fact they were used in spears. Clovis points are a type of flaked stone tool that can be made from rocks like flint or chert or volcanic glasses like obsidian. If these materials are struck in the correct way, flakes of rock can be chipped away to produce a sharp edge. By removing a series of small flakes, lumps of rock can be converted into a variety of tools. Clovis points can be distinguished from arrowheads and other ancient stone artifacts by the fact that they are rather large—often several inches long—and they have a characteristic "flute" at their base. This flute is the scar left by a flake removed from the rear end of the point. This flute, which may have been useful for attaching the point to a wooden shaft, is a very difficult feature to produce. If not done properly, the stresses applied to the base can easily snap the entire point in half. Today only the most skilled flintknappers can replicate Clovis points.

Clovis points have been found throughout the United States. At several sites, these artifacts have been found with the remains of mammoths, some have even been uncovered embedded in mammoth bones. These tools were therefore sometimes used as spear points by mammoth hunters, and since mammoths have not been around for a long time, they must be extremely old. Carbon-14 dates from sites associated with Clovis points confirm the antiquity of these tools, with the measured carbon-14 ages clustering around 11000 years BP. After calibration, we find that these artifacts date back to about 11000 BCE, or 13,000 years ago. These tools are consequently among the oldest evidence of human activity in North America. In fact, many archaeologists have argued that Clovis points could actually document the arrival of the first people to the New World.

There is absolutely no evidence that any of our prehuman ancestors, such as *Homo erectus*, ever made it into the Americas. Modern humans must therefore have found their way into the New World between 200,000 years ago, when they first appeared in Africa, and 13,000 years ago, when the Clovis

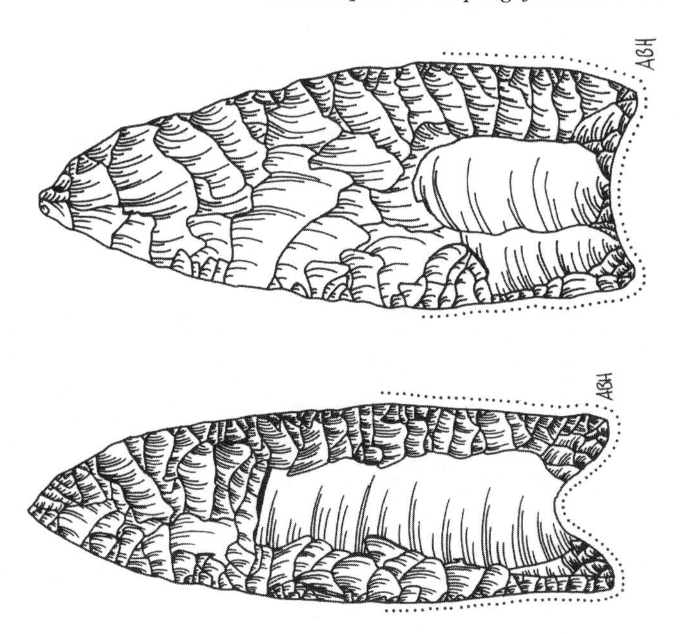

FIGURE 6.2 Examples of Clovis points. Note the distinctive "flute" at the base. Based on figures from John Whittaker *Flintknapping: Making and Understanding Stone Tools* (University of Texas Press, 1994).

points appear. It is most likely that they came via northeast Asia and Alaska, since this is where the Old and New Worlds are closest to each other, being separated only by the narrow body of water known as the Bering Strait.

This theory becomes even more plausible when we realize that there were at least two occasions when there was no water at all between Asia and Alaska. As we saw in the last chapter, shifts in oxygen isotope ratios and other climatic data indicate that the earth was significantly colder between 100,000 and 10,000 years ago. During this Ice Age, large amounts of water became locked up in huge glaciers and ice sheets, causing sea levels to drop. Around 50,000 years ago, and again around 20,000 years ago, the temperature was so low and sea levels fell so far that a land-bridge emerged between Asia and Alaska. Yet in spite of the ongoing Ice Age, large sections of Alaska and the land bridge were not covered by barren sheets of ice. This region was therefore able to support animals and people, and it was a natural conduit for human migrations into the New World.[1]

1. While other routes into the Americas have been suggested, they are very speculative and have not found wide acceptance.

While the ice age may have facilitated the arrival of *Homo sapiens* in Alaska, it also made traveling from Alaska to any other part of the Americas a challenge. Central Alaska was able to stay dry at this time because moist air moving in from the south and west dumped most of its snow when it passed over the coastal mountains, leaving the interior relatively free of snow or ice. By contrast, what is now Canada was almost completely covered in ice because the moist air from around the Gulf of Mexico produced snow as it moved north and east, enabling a gigantic sheet of ice to expand westward from Quebec and Baffin Island towards another ice sheet centered over the Rocky Mountains.

By around 35,000 years ago this ice would have blocked any overland route between Alaska and the rest of the continent. These glaciers began to melt about 20,000 years ago, and by about 14,000 years ago a corridor had opened up through western Canada into what is now the continental United States, as shown in Figure 6.1.

Clovis points from the western United States are almost as old as the formation of this ice-free corridor, hinting that the people who made the Clovis points might have used this route to enter this part of North America. Some archaeologists have even suggested that the makers of the Clovis points were the first humans to live south of the ice sheets. These people would have followed large animals like mammoths down the ice-free corridor into the plains of the western United States, and from there spread rapidly over the entire the entire New World following game, etc. When the large animals died out (possibly the result of the climatic changes brought on by the end of the Ice Age), the people began to settle down and use more local resourcesThis is almost certainly a much too simplistic picture of how people colonized the New World, but it is nonetheless consistent with the age, use, and distribution of the Clovis points. Furthermore, it is a useful model because it makes predictions that can be supported or challenged with additional data.

In particular, this model—sometimes called the Clovis-first model—suggests that there should not be any people in the Americas much before 13,000–14,000 years ago, since the ice-free corridor did not open until that time. Over the years, a number of archaeologists have claimed to discover sites in the Americas that predate the Clovis sites and to have disproved this model. None of these findings has been free of controversy, and since the better-supported claims all depend upon carbon-14 data, the debates usually center on the interpretation and reliability of these dates.

SECTION 6.2: MEADOWCROFT

The Meadowcroft rockshelter in western Pennsylvania is a shallow cave that has been occupied intermittently for millennia. For thousands of years, layer upon layer of rocks and dirt has accumulated in this cave, and some layers contain stone tools, campfires, and other indications of human activity. When this site was excavated in the late 1970s, charcoal from the fires in its lower levels had carbon-14 dates that suggested that the site was occupied over 15,000 years ago. This would put people in the Americas well before the ice-free corridor opened.

As with all other sites that seem to contain pre-Clovis artifacts, skeptics have challenged these findings. There are two issues that call Meadowcroft's age into question. First, the plant remains in the oldest layers of the site include the remains of oak and hickory trees, and it seems unlikely that such deciduous trees could survive here during the height of the ice age, when the ice sheet covering Canada was less than fifty miles away. The excavators counter this argument by saying the area around the site was sheltered and thus had a milder climate than one might expect.

The second issue with the age of the site has to do with the reliability of the carbon-14 dates themselves. Remember that western Pennsylvania is in the heart of coal country. The entire region is riddled with geological deposits containing remains of truly ancient—over 300 million years old—carbon-rich materials. All of the carbon-14 in these deposits has long since decayed away, so if any of this material mixed with the charcoal in the fires, it would dilute the carbon-14 fraction and the dates would be too old. The excavators contend that such contamination is not an issue, because the dates from different layers of the site are consistent. Put another way, the dates in any given layer are older than those of the layers above it and younger than the layers below it (see Figure 6.3). If the dates were altered, the contamination would have likely affected some layers more than others, causing the dates from different layers to be jumbled around. They also point out that microscopic examination of the dated samples showed no evidence of coal fragments or any other suspicious material.

While Meadowcroft remains one of the most promising candidates in North America for a pre-Clovis site, archaeologists continue to argue about its true age more than twenty-five years after it was first announced. These discussions often seem to degenerate into the sort of snarky bickering that I imagine most general-interest readers find more amusing or annoying than informative. The problem, by and large, is that the available data are very lim-

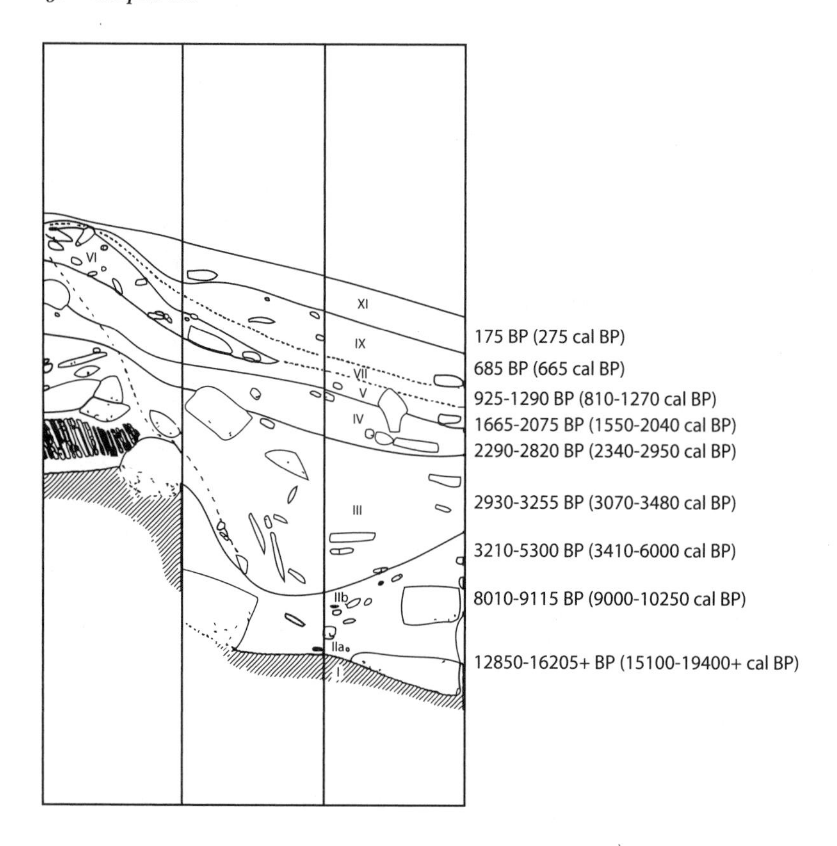

FIGURE 6.3 A cross-section of the Meadowcroft site, showing the various layers, along with the associated carbon-14 dates (based on J. M. Adovasio et al. "Meadowcroft Rockshelter 1977: An Overview" *American Antiquity* 43 (1978): 632–651). Note that deeper layers have older dates, as one would expect. The excavators present this as evidence that the dated samples were not contaminated.

ited, coming as they do from one particular site. In such a case, the excavators will typically have a deeper appreciation of the site and the specific challenges it presents. However, other archaeologists—quite naturally—do not want to rely on the excavators' word alone. Without data from other sites, the only way to evaluate the reliability of the excavators' claims is by looking critically at their methods and by reexamining the excavated material. Unfortunately, even if the excavators do not have the time, resources, or desire to answer every concern that a skeptic may have, they are still the only ones who can provide the requisite information, a situation that can easily lead to conflicts. The only solution to this problem is more data from more sites. While only a few sites have yet been found in North America with early dates even nearly as believable as those from Meadowcroft, there have been a number of very interesting discoveries from farther south.

SECTION 6.3: MONTE VERDE

When the Clovis-first picture was developed, South American sites were not nearly as well explored as those in the north. Since then, excavations and surveys have been providing a great deal of new data about the people who lived here near the end of the Ice Age. The most famous of these discoveries come from Monte Verde in south-central Chile. Unlike Meadowcroft, this site is located along the bank of a creek, and this stream exposed a layer of deposits a few feet below ground level containing stone tools and the well-preserved remains of both plants and animals. Archaeologists have found patterns in the distribution of these artifacts that suggested that people had once made a camp at this location. Specifically, they found a number of wooden poles and posts indicating the existence of a group of small huts or tents. Carbon-14 dates from this layer yielded dates around 12500 BP, or between 14,000 to 15,000 years ago after calibration. This site therefore appears to be somewhat older than Clovis and the ice-free corridor, which is particularly surprising given its location 10,000 miles south of Alaska.[2]

In 1997, the excavators of Monte Verde invited a group of independent archaeologists to their site to confirm the accuracy of their dates and interpretations. This team endorsed the antiquity of the site, making Monte Verde a popular candidate for a true pre-Clovis site. Of course, this action did not silence all criticism, and there have been several questions raised about the interpretation of the remains. As at Meadowcroft, the possibility of contamination is an issue, but in this case the primary concern is the relationship between the dated materials and the artifacts. Remember that the deposits were found near a stream. Such environments can collect all sorts of materials washed in from other locations, so it has been suggested that the artifacts found at Monte Verde do not come from a true human settlement, but instead are an agglomeration of material from a range of different times. The dates from charcoal and wooden objects at the site therefore do not provide a secure age of the clearly man-made objects like stone tools and pieces of twine. The excavators disagree with this assessment, pointing out that the dates of interest from the deposit generally fall within a relatively narrow range of time.

2. There are also hints of an even earlier occupation here, but they are extremely tenuous.

They also argue that the evidence for structures and anthropogenic patterns in the artifacts cannot be so easily dismissed.

SECTION 6.4: REGIONAL PATTERNS

Monte Verde renewed interest in alternatives to the Clovis-first model of the colonization of the New World. Right now, probably the most popular view is that while all of Canada was still covered in ice, some people were able to skirt the glaciers by traveling along the western coast of North America, using boats for at least part of their journey. While this scenario is certainly plausible, it is hard to find direct evidence for it or any other specific model of pre-Clovis migrations. Even assuming the early carbon-14 dates at places like Monte Verde and Meadowcroft are accurate, we still have only a handful of isolated pre-Clovis sites scattered over a broad region, which makes it very difficult to ascertain how people may have lived at this time. Furthermore, the available pre-Clovis sites have not yet provided many clues that would help archaeologists find additional early sites, frustrating many researchers and encouraging others' skepticism about the reliability of the early dates.

By contrast, since Clovis points are fairly elaborate and distinctive pieces of stonework, archaeologists can associate any site containing these artifacts with a common tool-making tradition. By combining data from a variety of different sites, researchers can develop and test ideas about how and when the people who made these points lived. For example, well-established carbon-14 dates from multiple Clovis sites in North America fall within a few hundred years of 11000 BCE, so we can be reasonably confident that most Clovis points were made and used during this limited period of time. Also, since these points are repeatedly associated with mammoth remains, mammoth hunting appears to have been an important activity for some of these people. By comparing the remains from these different sites, archaeologists can even deduce something about how they hunted and butchered such large animals. Of course, they probably did more than just hunt mammoths all day, but unraveling the other aspects of their culture is more difficult because we do not have a set of such distinctive, durable artifacts to study.

Similarly, I expect part of the reason the Clovis-first model has been so popular for so long was that it explained a pattern observed in multiple sites found in temporal and geographical proximity to one another, and that if patterns were discovered among non-Clovis early American sites, archaeologists would have a much clearer picture of the earliest cultures in the New World. For example, suppose that archaeologists were able to identify a regional pattern of sites connected with the earliest occupations at Meadowcroft. If any of these sites were found to be younger than the Clovis points or if they were found in locations that were under glaciers 15,000 years ago, then we would have some more justification for doubting the dates associated with this site. Alternatively, if these other sites were all located in areas *not* covered by ice back then, it would strengthen the case for the early occupation at Meadowcroft. Other patterns may even lend support to particular routes around the Canadian ice sheets. If people did bypass the glaciers along the coast, we might expect to have a collection of similar early sites scattered along the western coast of North America. Unfortunately, many of these coastal sites would have been drowned when sea levels rose after the end of the last Ice Age, making them particularly hard to find and study.

At present, the data from North America are still so sparse that it is difficult to discern clear patterns among the putative pre-Clovis sites. However, some interesting and unexpected patterns are beginning to emerge from the early South American sites. Recall that the Clovis-first theory suggests that the first Americans spread rapidly throughout the New World in pursuit of big game animals like mammoths. Even if people came to America earlier and through a different route, it is still possible that once there, they wandered far and wide hunting large animals. Such scenarios predict that early South American sites should be found in open environments that could support such large animals: the high Andean plains, for example. In fact, there is comparatively little evidence of early South American sites in upland areas. Instead, they are typically found along the coasts and in forested environments, including the lower Amazon. Remains found at these places indicate that these people were not primarily big game hunters. Instead, the evidence suggests that they hunted small animals, fished, collected shellfish, and gathered plants. For example, the possibly pre-Clovis and definitely early site of Monte Verde preserved potatoes, used packets of medicinal plants, and bits of cordage made from plant fibers. Another recently discovered site, Caverna da Pedra Pintada near Monte Alegre in eastern Brazil, preserves material that indicates people there lived on a diet of fruits and nuts in the middle of a tropical rain forest at almost the same time as people in North America were hunting mammoths with Clovis points. Indeed, the current carbon-14 data indicate that the North American sites containing Clovis points are not substantially earlier than non-Clovis sites like Caverna da Pedra Pintada. Instead, they are from essentially the same time period, which suggests that life in the New World 13,000 years ago was much more complex than the above models had predicted.

This new evidence for the diversity of lifestyles in the Americas at the end of the last Ice Age raises a host of questions for archaeologists. How were groups with different diets and habits distributed in the Americas over the centuries? Did certain groups specialize in specific resources? Were there any interactions between these groups and, if so, of what nature? It is impossible to address these questions fully with the available data. However, as more sites are discovered and additional age measurements become available, clearer patterns should emerge, enabling archaeologists to determine how people throughout the New World were using the land and its resources. Such research promises to eventually provide us with a much more complete picture of these earliest periods in American prehistory.

While these findings are just beginning to reshape our understanding of the earliest Americans, it appears that other new discoveries will soon cast light on a very different anthropological issue: how and when our ancestors first acquired the ability to walk upright on two legs. This phenomenon occurred millions of years before the appearance of modern humans, which means that the relevant material is so old that the samples do not contain sufficient carbon-14 to provide reliable age measurements. Instead, data obtained from another unstable isotope, together with information preserved in our genetic code, suggest that a breakthrough lies just around the corner.

SECTION 6.5: FURTHER READING

A good general work on North American archaeology is Brian Fagan *Ancient North America* (Thames and Hudson, 2000). Web-based resources on early American archaeology can be found through sites like those of the Center for the Study of the First Americans (www.centerfirstamericans.org) and the Paleoindian Database of the Americas (pidba.utk.edu).

A recent work discussing the Clovis points and what we know about the people who made them can be found in Gary Haynes *The Early Settlement of North America* (Cambridge University Press, 2004).

Detailed descriptions of Meadowcroft are to be found in R. C. Carlisle and J. M. Adovasio *Meadowcroft Rockshelter: Collected Papers on the Archaeology of Meadowcroft Rockshelter and the Cross Creek Drainage* (University of Pittsburgh, 1982). For a briefer description, see J. M. Adovasio et al. "Meadowcroft Rockshelter 1977: An Overview" *American Antiquity* 43 (1978): 632–651 or J. M. Adovasio et al. "The Meadowcroft Rockshelter Radiocarbon Chronology 1975–1990" *American Antiquity* 55 (1990): 348–354.

A description of the Monte Verde site is found in Thomas D. Dillehay *Monte Verde*, 2 vols. (Smithsonian Institution Press, 1989).

Skeptical perspectives on these sites (and associated "lively discussions") can be found in several places: the series of brief articles by Mead, Haynes, and Adovasio in *American Antiquity* 45, no. 3 (1980) (Jim I. Mead "Is It Really That Old? A Comment about the Meadowcroft Rockshelter 'Overview,'" pp. 579–582; C. Vance Haynes "Paleoindian Charcoal from Meadowcroft rock-shelter: Is Contamination a Problem?" pp. 582–587; and J. M. Adovasio et al. "Yes Virginia, It Really Is That Old: A Reply to Haynes and Mead," pp. 588–595); K. B. Tankersley and C. A. Munsun "Comments on the Meadowcroft Rockshelter: Radiocarbon Chronology and the Recognition of Coal Contaminants" *American Antiquity* 57 (1992): 321–326; "Monte Verde Revisited," a special report by *Scientific American, Discovering Archaeology*, November/December 1999, pp. 1–23; and Stuart J. Fiedel "Initial Human Colonization of the Americas: An Overview of the Issues and the Evidence" *Radiocarbon* 44 (2002): 407–436.

A good recent book on various aspects of the early inhabitants of the new world is: Nina Jablonski (ed.) *The First Americans* (University of California Press, 2002).

For more details about the various early South American sites, see A. C. Roosevelt, J. Douglas, and L. Brown "The Migrations and Adaptations of the First Americans: Clovis and Pre-Clovis Viewed from South America" in *The First Americans* edited by N. G. Jablonski (University of California Press, 2002) and Thomas D. Dillehay *The Settlement of the Americas* (Basic Books, 2000).

For a recent reevaluation of Clovis dates that indicates Clovis sites come from a very limited period of time and are contemporary with some non-Clovis sites in North and South America, see M. R. Waters and T. W. Stafford Jr. "Redefining the Age of Clovis: Implications for the Peopling of the Americas" *Science* 315 (2007): 1122–1126.

CHAPTER SEVEN

Potassium, Argon, DNA, and Walking Upright

From a strictly biological perspective, humans are not that different from chimpanzees or gorillas. We all have countless anatomical features in common—including opposable thumbs and fingernails—and our genetic blueprints differ only by a few percent. However, the small number of biological traits that do distinguish us from other apes correspond to enormous differences in behavior. A chimpanzee, after all, is unlikely to write a book like this one, and even if one did, it would have great difficulty finding a publisher. Therefore, if we can determine the origin of those characteristics unique to humankind, we can better understand and appreciate what it is that makes us special.

Two of the most obvious physical traits that distinguish humans from other great apes are our greatly enlarged brains and our style of walking on two legs with the torso held vertically. Our increased brain size clearly has a direct relationship to our uniquely complex behavior and culture. However, our posture also appears to have played a pivotal role in our evolution. Our ancestors walked on two legs long before they began to have bigger brains, and among all the ancestors of humans and great apes, only the bipedal creatures demonstrate dramatic increases in brain size. It is therefore possible that this peculiar mode of locomotion somehow facilitated later changes in brain structure, although the details of this relationship are far from clear.

The origin of our bipedalism is also a hot topic for anthropologists because it is a longstanding puzzle that recent discoveries may finally help solve. Changes in walking style, like changes in brain size, involve alterations of skeletal features, so in principle the fossil record should provide important information about when, where, and how both these crucial adaptations occurred. However, while ancient skulls from Africa document changes in brain size among our ancestors over the past five million years, no one has yet found old bones that clearly indicate when or where our ancestors began to walk upright. This lack of detailed information about the circumstances surrounding the origin of bipedalism has made it very difficult to determine why walking on two legs became advantageous for our ancestors. But in the last decade, teams of researchers working in Ethiopia, Kenya, and Chad have uncovered some very interesting fossils. At present, the recovered material is still rather fragmentary, but the dates associated with these finds are enough to make them extremely exciting to anthropologists. The ages of these newly discovered bones indicate these early bipeds lived at a time that—according to DNA evidence—may have been a crucial turning point in the history of our lineage, so these creatures may document the earliest stages of the unique traits like bipedalism that made us what we are today.

SECTION 7.1: THE HOMINIDS

The newly discovered fossils—like most fossils that anthropologists use to study the origins of our uniquely human traits—belong to a set of animals that all have some of the traits that make today's humans unique, such as enlarged brain size and bipedalism, as well as a host of subtler features including reduced canine teeth and barrel-shaped rib cages. These creatures used to be referred to as hominids, but thanks to recent refinements in the classification system they are now often called hominins instead. I will use the older, more familiar term here, but regardless of what you name them, the combinations of characteristics they share with us are unlikely to appear multiple times in different animals, so the hominids and modern humans almost certainly inherited these features from a common ancestor. The distribution of traits among hominid fossils can therefore reveal how and when our ancestors acquired these characteristics.

For example, consider brain size. Figure 7.1 shows the skulls of several different kinds of hominids from different times. The amount of skull rising over the eyebrow ridge provides a rough indication of brain size, so we can clearly see that the earliest hominids like *Australopithecus afarensis* have relatively small brains—comparable to those of modern chimpanzees. We can also observe a definite trend: brain size increases from *Australopithecus afarensis* to *Homo habilis* to *Homo erectus*, and finally to modern humans (also known as *Homo sapiens*). However, we can also see some hominids did not follow

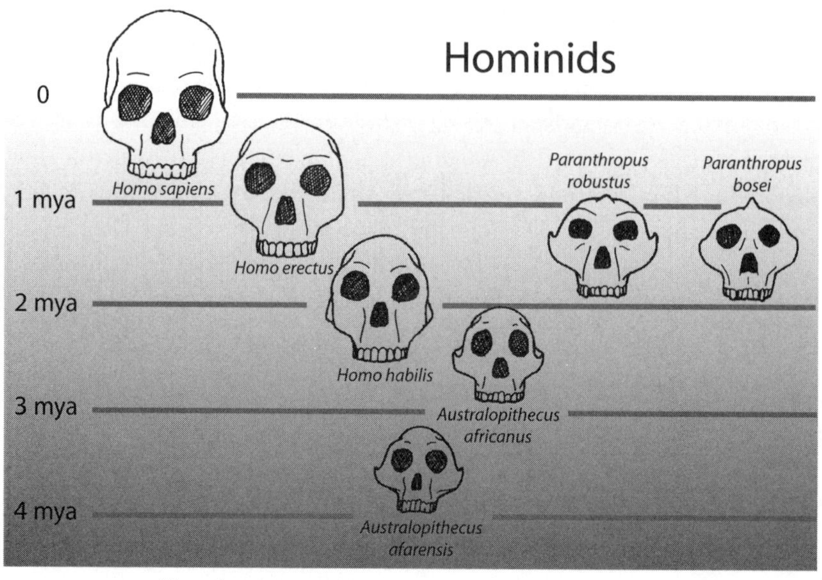

FIGURE 7.1 Drawings of various hominid skulls, all to scale. The position of the skulls along the vertical axis indicates the age of the fossil, while the horizontal axis is arbitrary. Increases in brain size can be observed as the dome of the skull rises over the top of the brow-line.

this same trend. For example, the brain size of the *Paranthropus robustus* is considerably less than that of the contemporary species *Homo erectus*. By comparing the habitats and diets of these different hominids, paleoanthropologists are able to gain insights into the processes that fostered the evolution of large brains. The data also indicate that hominids with enlarged brains began to appear about two million years ago, so the relevance of any climatic trends or other environmental phenomena at this time can be explored.

By contrast, all well-preserved, nearly complete hominid fossil specimens—even those from *Australopithecus afarensis*—have features that indicate these creatures walked on two legs: their lower spine was curved backwards in order to support a vertical trunk and their hip, knee, and ankle joints allowed their legs to swing forward and backward under the pelvis. This means that the ability to walk on two legs is not only older than enlarged brains, it is also older than *Australopithecus afarensis* or any of the other hominids shown in Figure

7.1. However, until someone discovers fossil hominids that could *not* walk efficiently on two legs, it will be very difficult to ascertain what caused our ancestors to adopt this mode of locomotion. Anthropologists seeking the origins of bipedalism have therefore been searching for fossil hominids predating *Australopithecus afarensis*, and these efforts have recently started to be rewarded.

In 2001, teams of paleoanthropologists working in Ethiopia and Kenya announced that they had found fragmentary remains of hominids. While only a few bones were found at each location, the characteristics of the teeth were sufficient to distinguish them from *Australopithecus* and other hominids, so they were given the names *Ardipithecus ramidus* and *Orrorin tugenensis*, respectively. More recently, a team digging in Chad found a well-preserved skull of yet another hominid, which they called *Sahelanthropus tchadensis*. These bones all appear to be older than any previously known hominid remains, and therefore document a previously unexplored period of hominid history. As with the early American sites described in the last chapter, the evidence for the antiquity of these new finds comes from a radiometric dating method based on an unstable isotope. However, in this case the isotope is not a form of carbon created high in the sky, but a form of potassium released from deep underground.

SECTION 7.2: POTASSIUM-ARGON DATING AND THE AGE OF HOMINID FOSSILS

Potassium atoms all have nineteen protons, and depending on the isotope, they can have varying numbers of neutrons. Most potassium on earth is in the form of potassium-39, which has twenty neutrons and is completely stable. However, about 0.01% of potassium atoms are in the form of potassium-40, which has twenty-one neutrons and is unstable (see Figure 7.2). Like carbon-14, this isotope of potassium can undergo beta decay by having a neutron spontaneously convert into a proton. In this case, the process leaves behind a calcium-40 nucleus. However, 10% of the time potassium-40 decays in a somewhat different way; the nucleus captures an electron, and one of the protons converts into a neutron, producing an atom of argon-40.

Like carbon-14, potassium-40 atoms can be used as timekeepers because they decay with a well-defined half-life determined solely by the number of protons and neutrons they contain. However, while carbon-14 has a half-life of only a few thousand years, potassium-40 has a half-life of 1.28 billion years, so these two isotopes probe very different timescales. If we have material that

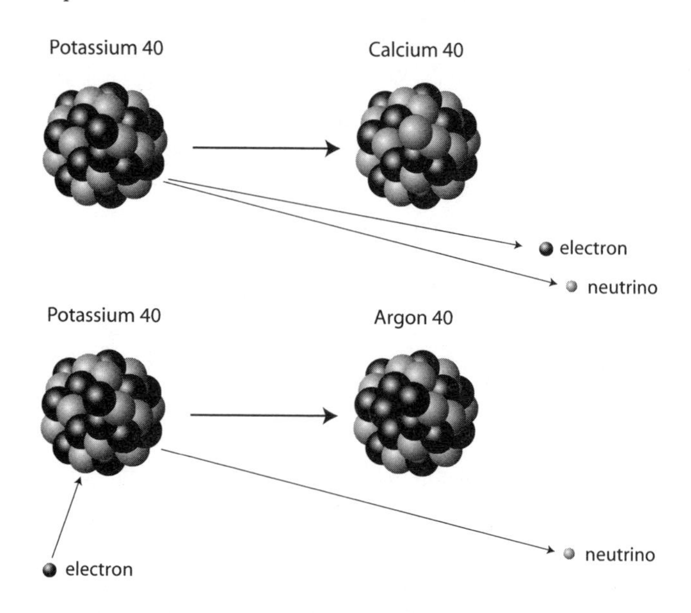

FIGURE 7.2 Potassium-40 decay. About 90 percent of the time it undergoes beta decay like carbon-14, and a neutron (black circle) converts into a proton (gray circle) to form calcium-40. The remaining 10 percent of the time the nucleus captures an electron, a proton converts into a neutron, and the nucleus converts into argon-40.

is some thousands of years old, enough time has gone by for a significant amount of the carbon-14 to have decayed, but very only a tiny fraction of potassium-40 has transformed into calcium or argon. For such comparatively recent material, carbon-14 provides a much more sensitive indicator of age than potassium-40. For objects millions or billions of years old, however, almost all of the original carbon-14 has decayed away while a significant amount of potassium-40 remains. Potassium-40 can therefore be used to measure the age of much older objects.

The types of materials that can be dated with these two methods are also very different because the relevant atoms have distinct chemical properties. As we have already seen, all organisms living at the same time receive comparable amounts of carbon-14 from the carbon dioxide in the atmosphere, so this isotope is often useful for dating material derived from living creatures. By contrast, potassium-40 is best used to date volcanic rocks, not because these all have similar potassium-40 contents, but because all molten rocks ideally contain no argon-40.

Argon belongs to a class of elements called noble gases, which includes both helium and neon. Noble gases are unique in that—except under very extreme circumstances—they do not form chemical bonds with other elements, so the only way they interact with other atoms is by bouncing off of them. Noble gases can easily escape from molten lava because these superheated rocks are in a quasi-liquid state, and their molecules are all moving and jostling past each other. In this environment, the argon atoms can just bounce around until they find the surface and hop off into the atmosphere. By contrast, argon can be trapped in a solid rock because here the atoms are arranged in rigid lattices, which form tiny cages that argon atoms cannot escape from.

Newly formed volcanic rocks ideally should not contain any argon since this gas escaped before the liquid lava cooled into a solid. However, they do typically contain at least some potassium-40. As time goes on, this potassium decays into argon-40, which remains trapped in the solid rock and allows us to estimate how much potassium-40 has decayed since the rock was formed. This data, combined with the current potassium-40 content of the rock, gives us all the information we need to compute the age of the rock.

For example, suppose we find a volcanic rock that currently contains 10 micrograms of potassium-40 and 1 microgram of argon-40. This means that 1 microgram of potassium-40 has transformed into argon-40 during the time since the rock first formed. Since only about 10% of potassium-40 atoms transform into argon-40, we can conclude that a total of 10 micrograms of potassium-40 has decayed over the course of the rock's existence. The rock therefore originally contained 20 micrograms of potassium-40 when it first cooled out of the lava flow, which means that one-half of the original potassium-40 atoms have decayed by the present day. This tells us that the rock must have solidified one potassium-40 half-life, or about 1.28 billion years, ago.

This method of measuring the age of volcanic rocks—called potassium-argon dating—is a simple and elegant way of measuring age because we can deduce the original potassium-40 content of the rock directly from the materials in the rock, and we do not need to estimate it through additional calibration data. In other words, the age estimate relies only on data provided by the rock itself and the only assumption made is that the rock initially contained no argon-40 at all. Still, this technique is certainly not foolproof, as there are a variety of processes that can contaminate the argon-40 content of a rock and corrupt the age estimate. The original lava flow may have contained some unmelted rocks, raising the argon-40 content of the lava above zero, or the rock may have been heated sometime after it formed, allowing some of the argon-40 to escape.

Just as with carbon-14 dating, there is a mini-industry dedicated to refining this technique and to developing procedures that provide accurate and reliable ages. For example, the reliability of potassium-argon dates can be evaluated using a clever bit of alchemy. Prior to extracting the argon, scientists can bombard the rock with neutrons from a nuclear reactor. Just as neutrons from cosmic rays convert nitrogen-14 into carbon-14 in the upper atmosphere, these neutrons transform some of the potassium-39 in the rock into argon-39. Since most of the potassium in any rock is in the form of potassium-39, this process generates a form of argon that "traces" the potassium content of the rock. After this treatment, the rock is gradually heated in order to extract the argon, and the argon-39 and argon-40 are separated and measured using mass spectrometry. Since both argon-39 and argon-40 were produced from the same element, the ratio of argon-39 to argon-40 should be the same throughout the rock. However, if the rock had been heated sometime earlier is its life and some of the argon-40 escaped, different regions or minerals in the rock will have discordant mixes of argon-39 and argon-40. Comparing the ratios of argon isotopes in the gas extracted from the rock at various temperatures therefore provides a way to determine whether the potassium-argon age has been corrupted.

SECTION 7.3: VOLCANOES AND FOSSILS IN EAST AFRICA

Potassium-argon dating allows us to determine when volcanic rocks first solidified. Since bones do not last long in the heat of a lava flow, we are not likely to find contemporary fossils embedded in these sorts of rocks. Most fossils are instead found in sedimentary deposits, where layer upon layer of mud, sand, or other material was piled up by water or wind. Normally, the potassium-argon method cannot help date sedimentary material directly. However, thanks to a geological phenomenon called the East African Rift System, the ages of volcanic rocks can help date sedimentary deposits containing early fossil hominids.

The East African Rift System is a place where the earth's crust has been torn apart by processes connected with those forces that caused the various continents to drift across the surface of the globe. These forces are driven by the heat contained deep within the earth, which is so intense that it makes rock pliable, allowing it to stretch and flow. Closer to the surface, rocks are cooler and more brittle, so instead of stretching, the uppermost crust breaks into pieces that slip past each other to form a series of valleys and depressions

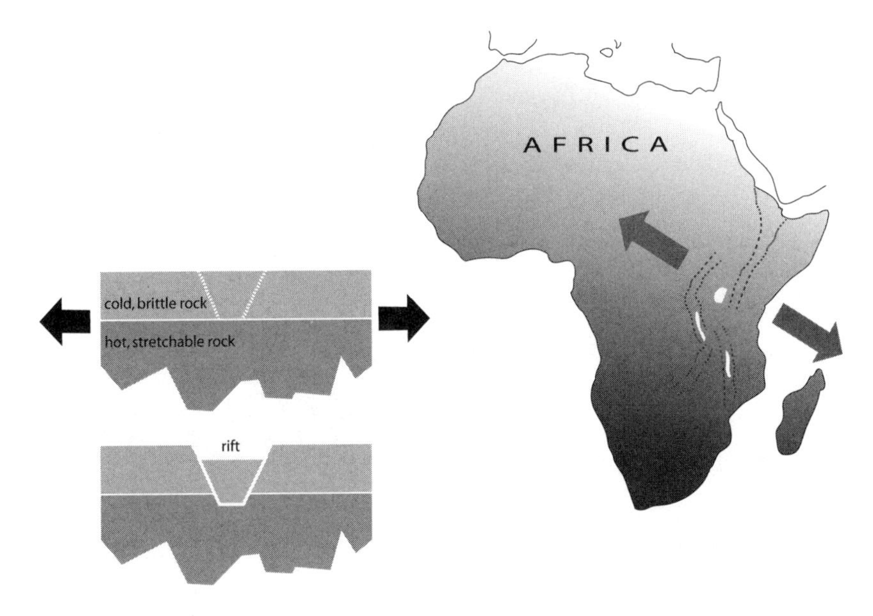

FIGURE 7.3 The East African Rift System. Left: a cross-section of a rift system. Geological forces pull the crust apart in these locations. The hot lower layer stretches, but the cooler upper layer breaks and a wedge slips down to form a depression. Right: the East African Rift System itself, with the depression indicated by dotted lines. The arrows indicate the forces that may have helped form these features.

(see Figure 7.3). This particular rift system extends all the way from Eritrea to Mozambique.

The East African Rift System had three important effects on the local environment. First, water collected at the bottom of the depressions, forming a series of lakes and creating habitats attractive to a variety of wildlife, including some hominids. Second, water and wind carried sediments down from the surrounding highlands into the depressions, which buried and preserved some of the animals as fossils. Third, the stress on and movement of the earth's crust allowed magma to reach the surface, causing widespread volcanic activity that at various times covered parts of this area with ash falls and lavas. East Africa therefore contains layers of volcanic deposits interleaved with fossil-bearing sediments.

Suppose a fossil-bearing layer of sedimentary rock is sandwiched between two layers of volcanic rocks. The fossil-bearing layer must be younger than

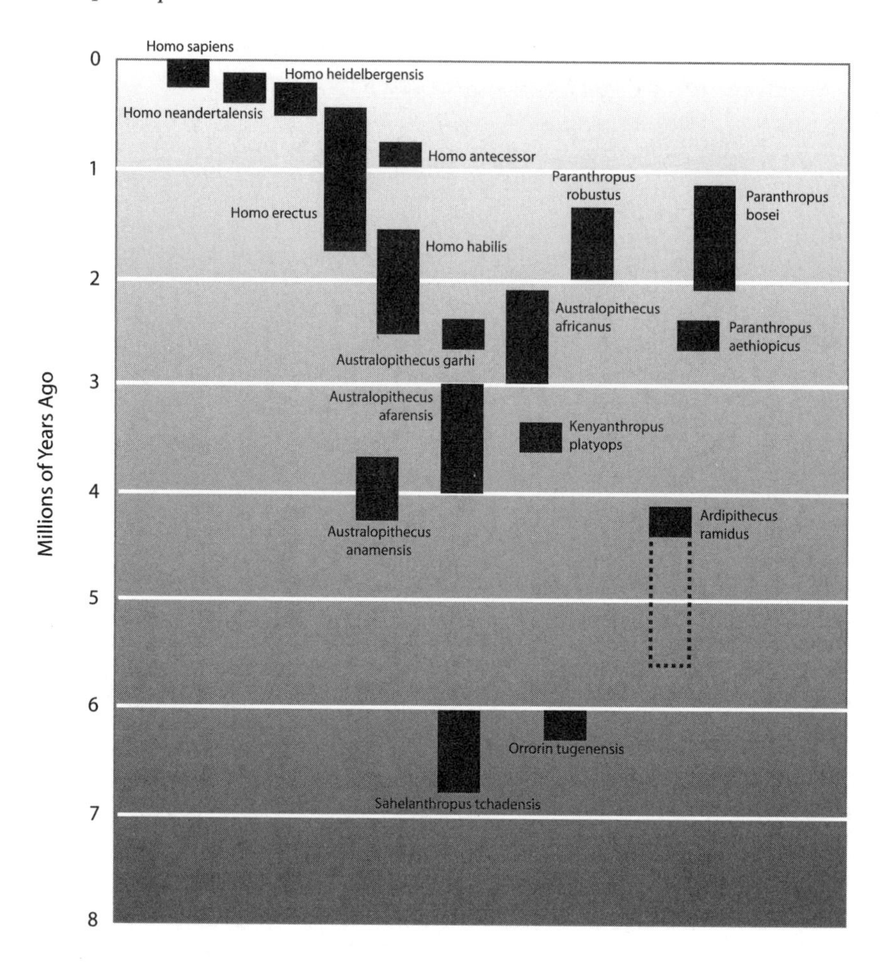

FIGURE 7.4 Dates of various different types of hominids, based on a figure in Bernard Wood "Hominid Revelations in Chad" *Nature* 418 (2002): 134–136. Bars indicate the range of time the various types of hominids probably lived. The recently discovered hominids *Ardipithecus*, *Orrorin*, and *Sahelanthropus* are significantly older than the previously known hominids, and probably come from the time when the ancestors of humans first began to walk upright.

the volcanic deposit it sits on and older than the volcanic rocks on top of it, so dating the volcanic layers with the potassium-argon system will provide tight constraints on the age of the fossil-bearing layer. This method has yielded very reliable age estimates for many of the hominid remains from the East African Rift System, including *Ardipithecus* and *Orrorin*. Even hominids found outside East Africa benefit from the rift system age measurements. For example, fossils belonging to *Sahelanthropus tchadensis* come from Chad, hundreds of miles west of the rift system and also far from any volcanic deposits that would facilitate dating. However, these fossils were found associated with the remains of other animals like the wild pig *Nyanzachoerus syrticus*, which are also found in the rift system. Using the method outlined above, paleontologists have deduced that these prehistoric beasts lived roughly six to seven million years ago. This suggests that the fossils, including *Sahelanthropus*, are from this same time period.

Figure 7.4 shows the ages of the hominids discovered as of 2006. Until about a decade ago, all known hominid remains were from deposits less than four million years old. The newly discovered fossils of *Ardipithecus, Orrorin,* and *Sahelanthropus*, however, date back to over six million years ago. These recently uncovered hominids are therefore much older than previous finds, but are they old enough to document the origins of bipedalism? The fragmentary remains include bones from the legs and feet of these creatures, and anthropologists are currently debating what the preserved material tells us about the posture of these early hominids. In spite of this, many anthropologists have high expectations for the material from this time period because studies of the DNA from humans and other primates indicate that these remains derive from a pivotal period in human evolution.

SECTION 7.4: MEASURING RELATIONSHIPS WITH DNA

Like the fossils found within the earth, the DNA within every living thing contains useful information about the history of life. As the methods of molecular biology continue to advance and improve, analyses of these molecules have come to play an increasingly important role in studies of the past. While there is still much work to be done before molecular data can even have a chance of providing age measurements that are as reliable as other methods, recent developments are very promising.

The double helix of deoxyribonucleic acid, or DNA, is a familiar icon in biology. These molecules exist in nearly every cell of our bodies and are composed of two intertwined spiral strands connected by a sequence of base pairs, each one made up of a pair of nucleotides. There are four different nucleotides in DNA: adenine, thymine, cytosine, and guanine, which are usually represented by the letters A, T, C, and G. The sequence of base pairs encodes information a cell needs to function and interact with other cells in a living organism. For example, certain parts of the sequence provide instructions for making different proteins, while other parts determine when these proteins should be made.

The nucleotides have specific chemical properties such that adenine and thymine always pair with one another across strands, as do cytosine and guanine. This means that if one strand has the series ACTTGCT, the other strand must have the sequence TGAACGA. Each of the two strands of the DNA molecule therefore contains essentially the same information. Normally this information is encased inside the coils of the double helix, but the machinery inside our cells can pull these two strands apart as needed so that the information inside can be read. Also, all of the data in a DNA molecule can be replicated by separating the two strands and using each one as a template for the construction of an identical copy of the original molecule. This process occurs naturally every time a cell divides, so that each of the cells maintains a complete set of the instructions required for it to function. By the same token, the information encoded in the DNA of every organism is derived from the DNA of its parents and is inherited by the DNA of its offspring.

Over time, as DNA is passed from generation to generation, the sequence of base pairs changes. These changes are called mutations, and they can occur due to errors in the replication process or because the DNA molecule itself gets damaged in some way. By changing the information encoded in the DNA, these mutations can change how certain cells function and ultimately alter the physical characteristics of the organism. Assuming that the change does not kill the cell or organism, the mutated DNA can be inherited by future generations of cells and organisms. Eventually, this mutated DNA will mutate again, and again that change can be passed on to new cells. As mutations in the DNA accumulate over many generations, creatures descended from a single organism can acquire very different characteristics. Indeed, it is likely that all living things on earth are descended from a common ancestor, and the diversity of life we see now is the result of huge numbers of mutations over billions of years.

Not only is the accumulation of mutations responsible for producing the great diversity of life on earth, it also enables us to uncover relationships between different organisms. Mutations are relatively rare, quasi-random events, and it is improbable—though by no means impossible—that the same mutation will occur twice in different organisms. Furthermore, after a mutation has occurred it is unlikely that an organism with that mutation will revert back to the exact same state as its premutation ancestors. This means that as mutations accumulate, the number of differences between DNA sequences tends to increase with time, and so two organisms with a recent common ancestor will have more similarities than two organisms with a more ancient common ancestor. We can therefore gain insight into the family history of organisms by studying their similarities and differences.

For many decades now, biologists have used the distribution of various physical characteristics to infer relationships between organisms and to investigate the patterns and processes behind the evolution of these traits. Nowadays, thanks to technology that can efficiently read the sequence of nucleotides on DNA molecules, biologists can compare long sequences of nucleotides from different creatures. These new data provide fascinating new insights into the relationships between organisms. In fact, these DNA sequences may provide a new, independent method of estimating how long ago related creatures—such as chimpanzees and humans—began to diverge from one another.

Variations in DNA sequences allow biologists to measure the differences between organisms in a more quantitative way than was previously possible. It is almost impossible to determine whether oak and elm trees are "more different" from each other than dogs and cats based solely on their appearances. Does the fact that cats always land on their feet while dogs don't make them *less* closely related than elm and oak trees, or do the different shapes of elm leaves and oak leaves mean that there are *more* generations separating them from each other than separate cats and dogs? The difficulties with this sort of inquiry are obvious. Within the DNA molecule, however, all of these differences boil down to the addition, removal, movement, and replacement of a discrete number of nucleotides. Therefore, it is possible to count the number of differences between species on the strands of their DNA. If you wanted, you could see how many times the DNA sequence of a cat had an A where a dog had a T, and then find the number of places an oak has an A where an elm has a T. By comparing these or any of a host of other possible parameters, we could actually make some quantitative statement about the differences between oaks and elms and cats and dogs. This quantitative data is essential for any attempt to estimate ages with biological data from modern animals.

The number of differences between two DNA sequences is a measure of how many mutations have occurred in the two sequences since they diverged from a common ancestor, so we expect that this number will get progressively larger as time goes on. Furthermore, if we assume that mutations accumulate at a steady rate, then this number also is proportional to the time that has elapsed since the two organisms last shared a common ancestor. For example, one stretch of DNA in polar bears differs by 1% from that found in grizzly bears, while the same sequence differs by 3% in wolves and coyotes. If the above assumption holds, we can deduce that the ancestors of the wolves and coyotes have had three times as long to accumulate mutations, so if polar bears and grizzly bears last shared a common ancestor about half a million years ago, we can estimate that the last common ancestor of wolves and coyotes lived sometime between one and two million years ago.

Living things are much more complicated than nuclear isotopes, so it is reasonable to question whether their mutations could ever accumulate at a fixed or even calculable rate. While it is true that there is still a lot of work that needs to be done before the reliability of this method can be assured, some of the available evidence is encouraging. Many animals use the same basic cellular machinery to read, repair, and copy their DNA molecules, so all of these creatures should be equally prone to mutations. Also, while exposure to certain toxic chemicals or large doses of radiation can greatly accelerate the mutation rate in organisms, such extreme conditions seldom occurred in the wilds of the distant past. We might therefore reasonably expect that mutations would occur at roughly the same rate in all organisms. We will explore this assumption in more detail in the next chapter, but here we must address a much bigger concern with molecular dating, which has to do with natural selection.

Mutations cannot accumulate unless they are passed on to another generation, but often the likelihood of this occurrence involves complex interactions between the organisms and their environment. For example, imagine a mutation that causes a rabbit to have a white coat instead of a dark coat. If the rabbit lived in a forest, it would stand out in the underbrush and promptly get eaten. Few or none of this hapless rabbit's offspring would survive to pass on this mutation. However, if it lived in the arctic, it would be well camouflaged and its offspring would likely thrive. These sorts of mutations will therefore be passed on through the generations at different rates depending on the particular situation. While such variations are of great interest to biologists, these mutations are clearly not going to be the most ideal timekeepers.

Fortunately, there are also mutations that are "silent," which means that they have no discernible impact on the creature's physical appearance or ability to survive. Silent mutations are possible because organisms do not typically use all of the information encoded in their DNA sequences. In fact, while we still do not know exactly what all the information encoded in any creature's DNA means, some mutations can be clearly identified as silent because much of the useful information is contained in segments of DNA with certain recognizable characteristics.

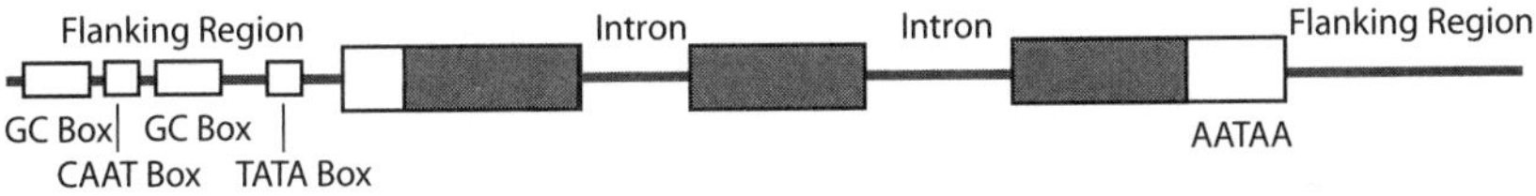

FIGURE 7.5 The general structure of a gene (based on a figure in Li's *Molecular Genetics*). Only the gray shaded regions contain the information for making the protein. Between these regions are the introns that are ignored when the protein is made. There are also flanking regions, which contain various characteristic sequences that allow the machinery in the cell to identify where the useful information is located.

For example, a significant portion of our DNA contains instructions for making various proteins, large molecules made up of long chains of chemicals called amino acids. Proteins are versatile molecules, and different proteins can have very different chemical properties depending on the sequence of amino acids they contain, so proteins are responsible for most of the complex chemical processes that allow a cell or an organism to function. The data required for making proteins are packaged in segments of DNA known as genes. Each gene contains a sequence of nucleotides that encode the sequence of amino acids required to make a particular protein. In order for the cell to be able to translate this information into a functional protein, it needs the flanking regions located on either side of the protein-coding sequence to tell it where the relevant information is. These regions contain characteristic nucleotide sequences (see Figure 7.5) that tell the relevant molecular machines where to start and stop reading the DNA.

Similarly, biologists use these flanking regions to identify all of the genes in a given stretch of DNA. It turns out that in humans and other mammals, only a small fraction of the DNA—roughly 5%—is in the form of functional genes. A portion of the remaining DNA still has some utility. For example, there are DNA sequences that are thought to regulate when the various genes are read. However, much of this material can be altered without having any noticeable affect on the cell or the organism. Some of this DNA has even been identified as "broken genes," former genes with mutations in the coding and flanking regions that render these DNA sequences impossible to read. Mutations in these regions should have no appreciable impact on the organism's health or appearance.

Even within the genes themselves, there are regions where changes in the DNA do not affect the structure of the protein. There are stretches of DNA called introns that are not used in the assembly of the protein. Furthermore, there are redundancies in the genetic code, so several different DNA sequences can correspond to the exact same sequence of amino acids. Many mutations within introns or between redundant sequences should therefore also be silent.

Since silent mutations do not affect how the organism interacts with its environment, both the probability that the mutation occurs and the probability that it gets passed on should not depend on where and how the creature lives. These mutations therefore are the ones most likely to accumulate at a steady rate and, by extension, to provide reasonable age estimates. Of course, the only way to evaluate the reliability of this technique is to look at real data from real organisms.

SECTION 7.5: PATTERNS IN THE MUTATIONS OF HUMANS AND APES

Humans and other primates have provided a useful test case for determining whether the accumulation rate of silent mutations can be stable enough to serve as a timekeeper. Their DNA has been studied for many years, and recently the molecular biologists Feng-Chi Chen and Wen-Hsiung Li published a paper on this subject. They took DNA from a human, a chimpanzee, a gorilla, and an orangutan and obtained the sequences of fifty-three silent or noncoding regions (for a total of 24,234 base pairs from each animal). They then looked for a specific type of mutation known as a point substitution mutation. These occur when a single base pair is replaced with another base pair, for example when the sequence ACTG becomes ACCG. These sorts of changes are quantified in terms of the fraction of nucleotides that differ between the two sequences. For example, the sequences:

ATTTCGCTAGCTAGTCGACGACTTCGATCAGCTAGCAGGCATCTGACGAGCT

and

ATATCGCTAGCTAGTCGACGACTTGGAGCAGCTAGCAGGAATCTGATGAGCT

have different nucleotides in five out of fifty positions, so the difference between these two sequences is 10%.[1]

1. In practice, the analysis is more complicated because some nucleotides may be missing from one of the sequences, which makes it more challenging to line up the two strings of letters properly. Also, we must account for the possibility that certain base pairs may have undergone more than one mutation.

TABLE 7.1 Genetic differences between humans, chimps, gorillas, and orangutans. Measured by Chen and Li.

human-chimp 1.24%	human-gorilla 1.62%	human-orangutan 3.08%
	chimp-gorilla 1.63%	chimp-orangutan 3.13%
		gorilla-orangutan 3.09%

With four animals, there are a total of six differences that can be measured: human-chimp, human-gorilla, human-orangutan, chimp-gorilla, chimp-orangutan, and gorilla-orangutan. Chen and Li calculated all of these differences (Table 7.1). From these six numbers, we can both deduce the relationships between these apes and argue that the mutation accumulation rate did not vary appreciably among the different animals.

Let us first reconstruct the relationships between these animals, assuming that they and their ancestors accumulated mutations at a steady rate. Relationships between animals are commonly depicted with a graph called a dendrogram or a phylogenetic tree, such as this:

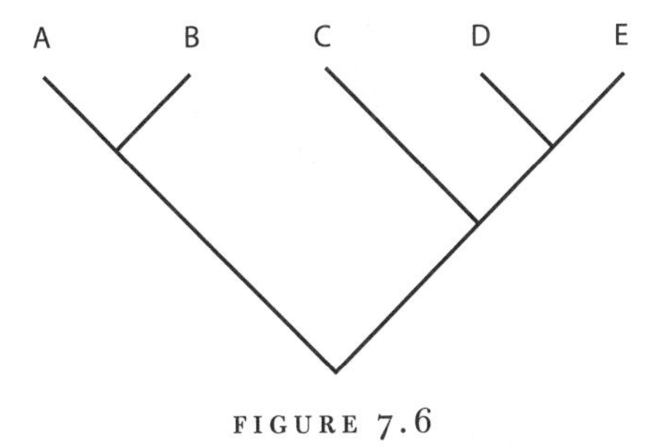

FIGURE 7.6

The letters at the top of the graph indicate a set of five animals living today, and the branching lines illustrate the ancestry of each of these organisms. In the recent past, all of them had distinctive ancestors, represented by the five separate lines leading to each letter. However, all of these creatures are ultimately derived from a single common ancestor, represented here by the point at the bottom of the plot. In between, the descendants of this common ancestor acquired mutations that set them apart from their relatives. In this case, the ancestors of creatures A and B diverged from the ancestors of creatures C, D, and E fairly early, while the ancestors of creatures D and E diverged from each other only recently.

We can begin to construct a phylogenetic tree for the great apes in Chen and Li's study by noting that the smallest of the six differences is between the human and the chimpanzee. This implies that these two animals have had the least amount of time to accumulate differences, so the ancestors of chimps and humans must have diverged after any split involving gorillas or orangutans. We can represent their relationship like this:

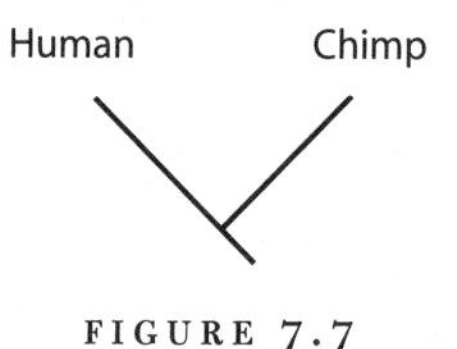

FIGURE 7.7

Next, we observe that both chimps and humans have fewer differences with gorillas (about 1.6%) than they do with orangutans (roughly 3.1%), so the divergence between the ancestors of chimpanzees and humans and the ancestors of gorillas occurred more recently than the split with the ancestors of orangutans. Humans, chimps, and gorillas therefore share a common ancestry illustrated by the following tree:

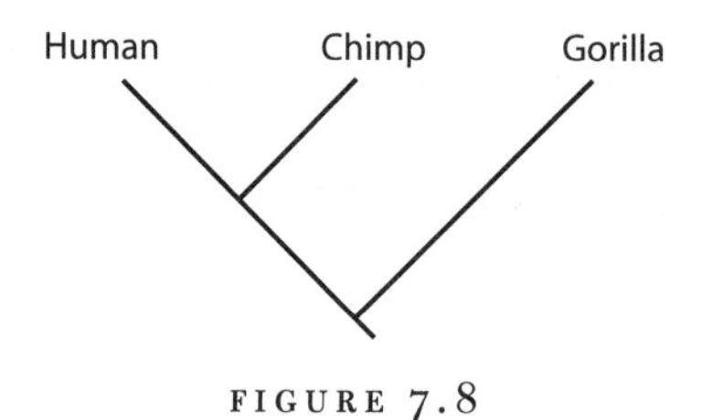

FIGURE 7.8

This leaves the orangutans, which are the most distinctive animals in this study. The ancestors of orangutans must have been accumulating distinct mutations for the longest period of time, so they are the first group to branch away from the common ancestor of all of these animals, as illustrated in the completed phylogenetic tree:

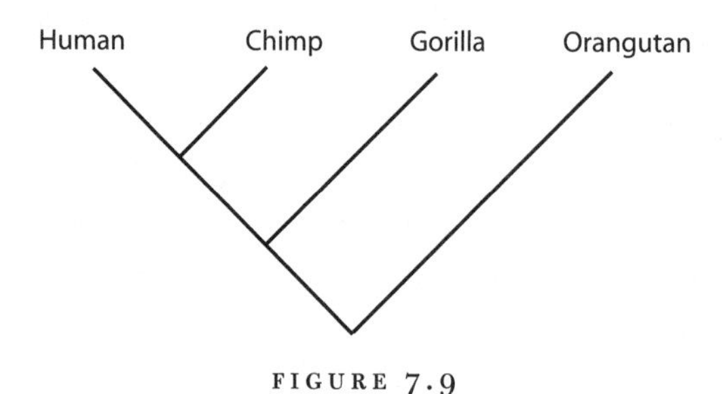

FIGURE 7.9

This graph encodes a brief history of the great apes, which starts with a species of primate that would ultimately give rise humans, chimps, gorilla, and orangutans. At some point in the past, some descendants of this group began to acquire the mutations that would ultimately become the distinguishing traits of orangutans, such as extremely close-set eye sockets. Meanwhile, a different group of descendants acquired mutations that produced the characteristics seen in gorillas, chimpanzees, and humans. Later, this second group of animals itself broke into two groups, one that gives rise to gorillas and another that formed the ancestors of chimps and humans. Finally, the ancestors of humans and chimps diverge, and each acquires distinctive mutations. While it is likely that there were many other branching events similar to the three indicated in this graph, none of these events produced descendants that have survived to the present day.

The above diagram was generated assuming that all of these animals accumulated mutations at the same rate. We can check whether this was in fact the case by taking a closer look at the measured differences. If the ancestors of humans accumulated mutations *faster* than the ancestors of chimpanzees, then the measured difference between humans and gorillas should be *larger* than the difference between chimpanzees and gorillas, but this is not consistent with the data. The data instead indicate that the difference between humans and gorillas is almost identical to the difference between chimpanzees and gorillas, so the ancestors of humans and chimpanzees seem to have been accumulating mutations at nearly the same rate. Similarly, since gorillas, chimpanzees, and humans all differ from the orangutan by about 3.1%, the ancestors of all of these animals also appear to have acquired mutations at the same rate. Although this does not absolutely prove that the mutation accumulation rate was constant, it does tend to support the idea. Otherwise, it would be quite a coincidence if three different primates in three different environments all managed to accumulate an equivalent number of mutations over the same time period. More recent research has found that the mutation rates among the ancestors of the great apes did differ slightly, but the variations are sufficiently small (at most about 10%) that we can ignore them here.

Given that the great apes do appear to have accumulated mutations at an approximately constant rate, we can attempt to use these data to estimate when the various divisions actually occurred and, consequently, when our ancestors first acquired the ability to walk upright. We have already seen that since the number of differences between chimps and humans is smaller than the number of differences between orangutans and humans, the ancestors of chimps and humans must have parted company more recently than the ancestors of humans and orangutans. Now we will be more specific. The difference between chimps and humans (1.24%) is roughly two-fifths the difference between orangutans and humans (3.08%), so the chimp-human split must be that much more recent than the orangutan split. Integrating this additional information into the tree, we get this diagram:

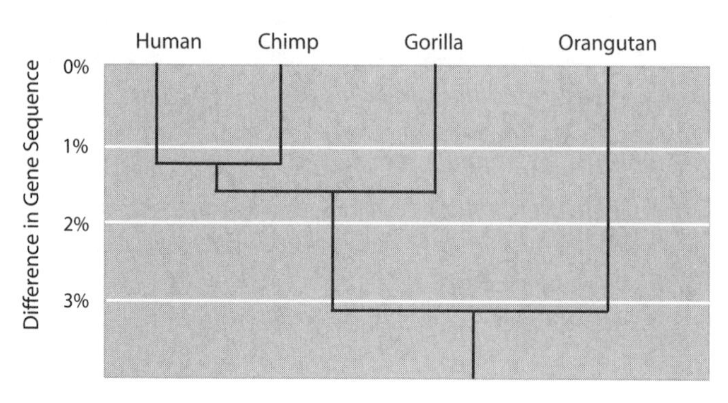

FIGURE 7.10

Here, the horizontal lines represent equal units of time, and the positions of the various branchings indicate when these events occurred. On the left are the percentage differences in the gene sequences. In this figure, the orangutan line splits off just before the time corresponding to 3% difference, and the human and chimpanzee lines split somewhat before the time that results in a 1% difference. Of course, we still need to figure out how many years it takes to accumulate a given variation. In the future, it may be possible to calculate the relevant timescales based on the molecular biology of the relevant organisms, but not yet. For now, mutation accumulation rates are estimated based on the fossil record.

While the fossil record of humans, chimps, and gorillas still does not provide enough information to indicate when exactly their ancestors diverged from the other apes, orangutans are another story. Paleontologists have found the fossils of an animal named *Sivapithecus*, whose skull has many features—such as close-set eye sockets and incisors of varying sizes—that among great apes are now seen only in orangutans. These similarities indicate that *Sivapithecus* is derived from the same line of animals that produced modern orangutans. These fossils have been found in deposits dating to around twelve million years ago, so the ancestors of orangutans must have started to acquire their distinctive traits sometime *earlier* than this.

Further information on the origin of orangutans has been gleaned from a fossil animal named *Proconsul*. This animal has features that are shared by all of the great apes—for example, it lacks a tail—but it has none of the features that are unique to chimps, humans, gorillas, or orangutans. Therefore, this creature probably existed before any of the living apes' ancestors acquired distinguishing characteristics. This beast is found in deposits from twenty million years ago, so ancestors of the orangutans probably acquired their distinctive characteristics sometime *after* this.

Together, the existence of *Sivapithecus* twelve million years ago and *Proconsul* twenty million years ago strongly suggests that the ancestors of the orangutans first diverged from the ancestors of the other apes about sixteen million years ago, give or take a few million years. Since it took sixteen million years for orangutans and other apes to acquire differences of 3.1% in these DNA sequences, the differences between two lines of great apes must increase at a rate of about 1% per five million years, allowing us at last to put a proper timescale on our family tree of the great apes:

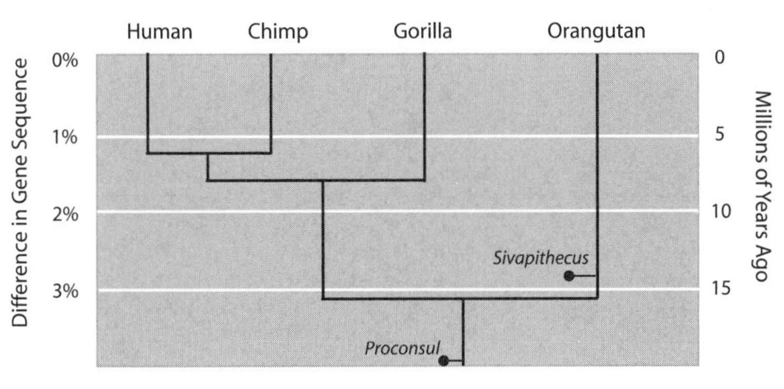

FIGURE 7.11

In addition, we can estimate that the lines leading to humans and chimps—which today differ by about 1.24%—began to diverge roughly 6.5 million years ago. This sort of analysis has been done many times using similar methods but different combinations of primates and fossil calibrators, and the results are usually the same.

Together, the molecular and the fossil evidence suggest that anthropologists are on the threshold of making a very exciting discovery. The fossils of *Australopithecus* demonstrate that our ancestors were already able to walk on two legs 4.5 million years ago. Meanwhile, the molecular data indicate that our ancestors began to acquire distinctively human like traits—like bipedalism—only between six and seven million years ago. The first hominids to habitually walk on two legs therefore should have lived around five to six million years ago.

The newly discovered fossils of *Ardipithecus*, *Orrorin,* and *Sahelanthropus* date from this critical time period. Thus far, the available material is still too fragmentary to clearly document the earliest stages of hominid bipedalism, but as more and more remains of these and other similarly ancient hominids are found, they will hopefully document when, where, and how our ancestors began to walk upright. Such findings would be a boon to the study of human and primate evolution and also validate molecular methods of measuring the passage of time. Of course, more complete fossils of these animals could also surprise us and perhaps cause us to reevaluate our relationships with the other apes. In either case, the future for this field should be very interesting.

Beyond the study of human origins, new genetic sequence data is also having a major impact on many other areas of biology and paleontology. For instance, molecular data may provide important clues about the interrelationships and origins of modern groups of mammals like bats, rodents, primates, and whales. As we will see in the next chapter, this sort of research requires much more sophisticated analytical techniques to cope with the large amount of DNA sequence data involved as well as the comparatively large divergences between organisms. The reliability of these methods is somewhat uncertain at the moment, but they can still provide intriguing data and may eventually yield a clearer picture of mammalian evolution during the end of the age of dinosaurs.

SECTION 7.6: FURTHER READING

For information on human evolution at a popular level, try Ian Tattersal and Jeffrey Schwartz *Extinct Humans* (Westview Press, 2000) and Carl Zimmer *Smithsonian Intimate Guide to Human Origins* (Smithsonian Books, 2005). A good source on the web with many links is www.talkorigins.org/faqs/homs/. For more details, try Glenn C. Conroy *Reconstructing Human Origins* (W. W. Norton, 1997).

For detailed information about the recently discovered ancient hominids, see the news article B. Wood "Hominid Revelations from Chad" *Nature* 418 (2002): 133–136. Subsequent discussions of these early hominids can be found in K. Galik et al. "External and Internal Morphology of the BAR 1002′00 *Orrorin tugenensis* Femur" *Science* 305 (2004): 1450–1453, and T. D. White et al. "Asa Issie, Aramis, and the origin of *Australopithecus*" *Nature* 440 (2006): 883–889.

For potassium-argon dating, see R. E. Taylor and M. J. Aitken *Chronometric Dating in Archaeology* (Plenum Press, 1997), chapter 4. A nice treatment can also be found in geology textbooks like Brian J. Skinner and Stephen C. Porter *The Dynamic Earth,* 2nd ed. (John Wiley and Sons, 1992).

For the basics of genetics, a good place to start is Larry Gonick and Mark Wheeler *The Cartoon Guide to Genetics* (Perennial Press, 1991).

Some books on reconstructing relationships from genetic data at the college level are Wen-Hsiung Li *Molecular Evolution* (Sinauer, 1997) and M. Nei and S. Kumar *Molecular Evolution and Phylogenetics* (Oxford University Press, 2000).

For the details of the genetic analysis cited in this chapter, see Feng-Chi Chen and Wen-Hsiung Li "Genomic Divergences between Humans and Other Hominids" *American Journal of Human Genetics* 68 (2001): 444–456. For a more recent analysis of great ape DNA, see Navin Elango et al. "Variable molecular clocks in the hominoids" *Proceedings of the National Academy of Sciences* 103 (2006): 1370–1375.

CHAPTER EIGHT

Molecular Dating and the Many Different Types of Mammals

About sixty-five million years ago, a series of events unfolded that culminated in some of the most dramatic changes in the history of life on earth. Massive volcanic activity on what would become India probably released large quantities of toxic gases into the atmosphere, and a meteor about ten kilometers in diameter crashed into the northern edge of the Yucatan peninsula, spreading devastation far and wide. These catastrophic events, perhaps in concert with more long-term processes involving diseases or climate change, put significant stress on the biosphere. During this trying time, the last of the large dinosaurs like *Triceratops* and *Tyrannosaurus* went extinct, along with pterosaurs, mosasaurs, plesiosaurs, and a host of other creatures.

While this marked the end of the age of giant dinosaurs, it was the beginning of a new era for mammals. Prior to the events of sixty-five million years ago, mammals were generally small, shrew-like creatures, although a few of them were as big as medium-sized dogs. Afterwards, they became important parts of many ecosystems, and different groups of mammals acquired a wide variety of specialized traits, including hooves, paws, flippers, and even wings. The fossil evidence reveals that much of this great increase in diversity happened relatively rapidly, within about ten million years of the meteor impact. The mammals that survived the meteor impact probably spread into multiple environments, and as different groups adapted to different situations, they evolved a variety of different characteristics.

This formative era in the evolution of mammals promises to provide many insights into how groups of animals respond to changes in the world's ecosystems. However, the very changes that make this time period so interesting also make it challenging to study. The fossil record only preserves the remains of some of the creatures that lived in any given place at any given time. The distinctive physical characteristics in the fossil remains often provide sufficient information to document the relationships between creatures from different deposits, allowing paleontologists to trace the history of different groups of animals. However, there are few traits that can be used to link the somewhat shrew-like mammals from the end of the age of the dinosaurs with specific groups of later mammals like rodents or primates, so the relationships between these early mammals are rather uncertain. While paleontologists continue to uncover fossils from this important time period, other biologists are attempting to trace the origins and history of modern mammals by using the DNA from living animals to estimate the ages of various branches of the mammalian family tree.

SECTION 8.1: THE DIVERSITY OF MAMMALS

Before we begin to investigate the history of mammals, we must first familiarize ourselves with the different kinds of mammals alive today. There are thousands of different species of living mammals, which come in wide variety of sizes and shapes: cats, dogs, people, bears, armadillos, bats, rats, mice, guinea pigs, deer, cows, hippos, whales, horses, anteaters, hedgehogs, monkeys, manatees, aardvarks, just to name a few. In spite of this bewildering diversity, all of these creatures have several very important traits in common. They all have hair or fur at some point in their lives, they are able to regulate their body temperature, and they can nourish their young with milk.

To modern biologists, these and other commonalities do not just provide a way to distinguish mammals from other organisms; they also reveal something about the ancestry of these beasts. Characteristics like furry coats arose due to a series of mutations in the genetic code of various organisms that were passed on to their offspring. If multiple animals have such a trait, then either they all inherited this trait from a common ancestor who had that characteristic, or the mutations responsible for the trait happened more than once in different organisms. Since mutations occur quasi-randomly throughout the DNA of an organism, the chances that the specific mutations responsible for a particular trait will occur more than once is small. Shared features due to mutations therefore are suggestive of a common ancestry.

While common traits can provide evidence of the relationships between organisms, we must be careful to account for the possibility of convergence. If distantly related organisms are placed in similar environments, then similar traits may be advantageous to the survival of both those organisms and their descendants. For example, both a rabbit and a fox living in the arctic would find a white coat preferable during the winter. This means that mutations that produce a white winter coat are disproportionately likely to be passed on through the generations among the descendants of either the rabbit or the fox. The descendants of both these animals could therefore easily wind up with white coats, not because they were all descended from a single animal with white fur, but instead because that trait was favorable for both foxes and rabbits. To distinguish traits derived from a common ancestor from those due to convergence, biologists must carefully study many features and their distributions between different organisms.

Beyond their hairy coverings, ability to produce milk, and high metabolisms, all mammals share a multitude of more subtle features that set them apart from other animals: the arrangement of three tiny bones in our middle ears is not found in birds or reptiles, and the teeth of mammals are in general much more complex than those of other creatures. All of this is strong evidence that all mammals share a particular common ancestor and that we are all more closely related to each other than any one of us is to any other animal.

Besides showing that all mammals are related, a detailed study of physical characteristics also provides a way to categorize these creatures. All of today's mammals can be divided into three broad groups: monotremes, marsupials and placentals. Monotremes are mammals that lay eggs, such as the platypus or the echidna. Marsupials, like kangaroos and koala bears, give birth to live young which develop inside their mother's pouch. Placental mammals, which account for about 90% of living mammals, bear live young and use a placenta to support the fetus during gestation.

In this chapter, we focus exclusively on the placental mammals, which are subdivided into roughly twenty groups called orders (Table 8.1). The names of some of these groups may seem odd at first, but this classification scheme is not nearly as esoteric as the nomenclature might suggest. The beasts in each order share a number of features that clearly distinguish them from other mammals and suggest that they belong to the same branch of the mammalian family tree.

One of the most obvious subsets of placental mammals is the bats, the only mammals that can actively fly. All bats have wings formed by a membrane that is supported by their highly elongated fingers. No other mammal has wings like this, and this feature alone makes it almost impossible to mistake a bat for any other type of mammal. A more detailed study of their anatomy reveals additional features unique to bats, many of which are adaptations to

TABLE 8.1 Orders of placental mammals except for Insectivora (see Table 8.2). Numbers of species based on D. E. Wilson and D. M. Reeder *Mammals of the World*, 2nd ed. (Smithsonian Institution Press, 1993).

Order	Number of species	Includes
Rodentia	1995	mice, squirrels, guinea pigs
Lagomorpha	80	rabbits, hares
Chiroptera	925	bats
Carnivora	280	bears, cats, dogs, weasels, seals
Primates	233	monkeys, apes, humans
Cetacea	78	whales, dolphins
Artiodactyla	215	cows, pigs, llamas, deer, sheep
Perissodactyla	18	horses, rhinos, tapirs
Xenarthra	29	sloths, anteaters, armadillos
Pholidota	7	pangolins
Dermoptera	2	flying lemurs
Tubulidentata	1	aardvarks
Sirenia	4	manatees, dugongs
Hyracoidea	11	hyraxes
Proboscidea	2	elephants
Macroscelidea	15	elephant shrews
Scandentia	19	tree shrews

powered flight. Bats are therefore understandably placed in their own order Chiroptera.

Another familiar grouping of mammals consists of the whales and dolphins. These mammals spend all their lives in the water. They have streamlined bodies, long tails with horizontal "flukes" on the end, no external hind limbs and forelimbs shaped like flippers. Like the bats, these mammals are given their own order, named Cetacea.

Certain groups of mammals are most easily identified by looking at their teeth. Many mammals, like mice, rats, squirrels, and guinea pigs, have the same type of oversized incisors, one pair each in their upper and lower jaws. These teeth grow throughout their life and acquire a chisel-like edge through constant wear. Mammals with teeth such as these are grouped together in the order Rodentia. Other mammals, such as rabbits, have a similar set of continuously growing incisors, but also have two ever-growing peg-like teeth in their upper jaw that are not present in rodents. This distinction, among others, is why rabbits and hares are not classified as rodents, but are instead placed in the separate order Lagomorpha.

Another group of mammals with distinctive teeth includes cats, dogs, bears, weasels, and seals. Of course, the most obvious teeth in many of these animals are their fangs. However, similarly large canines are found in baboons and other mammals, so while impressive, these teeth are not that useful for classification. Instead, we must look further back in the mouth, to the last premolar in the upper jaw and the first molar in the lower jaw, which are specialized to act like a pair of shears. The animals with this trait are placed in the order Carnivora, so-named because most of its members are carnivorous.

Instead of distinctive teeth, some groups of mammals have distinctive hands and feet. For instance, monkeys, lemurs, and apes (including humans) have opposable digits, and have nails instead of claws on at least their big toes. We also tend to have large brains and large eyes. These traits allow these mammals to be placed together in the order Primates.

Another distinctive type of feet is the hooves found on horses and cows. In fact, there are two very different types of hoofed animals with very different structures in their feet. In many hoofed animals, including, cows, sheep, deer, and pigs, the axis of the foot runs between two toes, so these animals tend to have an even number of hooves. These cloven-hoofed animals are grouped together in the order Artiodactyla. On the other hand, in horses and rhinos, the axis of the foot runs along one toe, and these animals tend to have one or three toes. This requires a significantly different foot and ankle structure, so these odd-toed hoofed animals are placed in a separate order, Perissodactyla.

One group of mammals is distinguished not by their teeth or their toes, but by their spines. Anteaters, armadillos, and sloths are all rather strange-looking creatures that come mainly from South America. All of these animals have a set of distinctive processes in the vertebrae of their lower back that are found in no other placental mammals. They are consequently placed in their own order, called Xenarthra.

Certain types of animals are so distinctive that they are placed their own orders. Elephants, for example, are the only living examples of the order Proboscidea.

Some mammals have some superficial similarities with one of the above groups, but on closer inspection they are found to lack the diagnostic features of those groups. For example, pangolins ("scaly anteaters") and aardvarks look a lot like anteaters, and indeed all three animals dine mainly on ants and termites. However, neither pangolins nor aardvarks have distinctive xenarthran processes in their vertebrae. This means that neither of these animals can be placed in the order Xenarthra. Furthermore, each of these animals has its own unique traits. Pangolins, for example, are covered with large "scales"

TABLE 8.2 Families historically included in order Insectivora. Numbers of species based on D. E. Wilson and D. M. Reeder *Mammals of the World*, 2nd ed. (Smithsonian Institution Press, 1993).

Family	Number of species	Includes
Erinaceidae	21	hedgehogs
Talpidae	42	moles
Soricidae	312	most shrews
Tenrecidae	24	tenrecs
Chrysochloridae	18	golden moles
Solenodontidae	2	solenodons

that make them look like pine cones. These animals are consequently given their own orders: Pholidota for pangolins and Tubulidentata for aardvarks.

In a similar way, hyraxes look a little like cat-sized rodents, but these mammals have four incisors in their lower jaw and hoof-like nails not found in any rodent, so the separate order Hyracoidea was created for them. Manatees and dugongs, on the other hand, are fully aquatic and have a basic body shape similar to whales. However, they are strictly herbivorous and have a distinctive snout and teeth adapted specifically to graze on underwater vegetation. They are placed in the order Sirenia. Finally, there are the so-called flying lemurs, a double misnomer because these animals are not lemurs, nor can they truly fly. They instead glide on a furry membrane extending between their front and back legs. They also have strange front teeth shaped like little combs. These unusual animals are assigned to the order Dermoptera.

There are several hundred species of placental mammals that are not included in any of these orders. These are mostly small creatures such as shrews, moles, and hedgehogs. All of these animals were once placed in a single order Insectivora, because most of them ate insects. However, there are no distinctive characteristics common to all of the members of this group, and it had long been recognized that this "order" might just be a catch-all term for several distinct groups of animals. Indeed, the elephant shrews, which have a long, flexible snout, and the squirrel-like tree shrews are now generally accepted to belong to their own distinct orders, named Macroscelidea and Scandentia respectively.

More recently, the unity of the remaining "insectivores" has come under question. Table 8.2 lists the six families of mammals typically included within the Insectivores. The family Erinaceidae includes hedgehogs, while the family Talpidae includes moles. The family Soricidae contains most types of shrews. The family Tenrecidae is a diverse group of mammals confined to Madagascar and Africa, which can superficially resemble shrews, moles, hedgehogs, and even otters. Similarly, the Chrysochloridae or African golden moles are burrowing animals like moles, but their front legs stay under their body while they dig, unlike moles—whose front legs stick out sideways. Finally, the family Solenodontidae consists of two species of shrew-like Solenodons found on Cuba and Hispaniola.

SECTION 8.2: ARRANGING THE ORDERS WITH MORPHOLOGY AND MOLECULES

The similarities within each of the above orders (except perhaps Insectivora) suggest that the mammals belonging to one order share a common ancestry. In other words, each order corresponds to a different branch of the mammalian family tree. However, identifying the relationships *between* these different groups has been challenging. While each order has its own unique characteristics, there are few informative traits that are shared across multiple orders. This makes it difficult to ascertain, for example, whether bats are more closely related to rodents, primates, or shrews. Even though paleontologists have found fossils representing all of the different orders (see Figure 8.1), at present few of these fossils shed light on the relationships between different orders. For example, the earliest known bat fossils are of creatures that already had fully functional wings. These, of course, do not provide much information about how or when bats acquired their ability to fly.

One important exception this situation involves the ancestors of whales. Skeletons have recently been found that belong to early ancestors of whales. These fossils were found in areas of Pakistan that used to be either lakes or part of an ancient sea, and their limb proportions confirm that these proto-whales were indeed semi-aquatic. However, much unlike modern whales, these beasts had substantial hind legs. The anklebones of these creatures include a bone called the astragalus with a distinctive "double-pulley" shape, which allows bones on both sides of it to rotate back and forth. This distinctive anklebone is a trait now found only in artiodactyls like cows, pigs, and hippos. This suggests that in spite of the many patently obvious differences today, cetaceans are in fact closely related to artiodactyls.

After many more discoveries of this sort, we may eventually be able to clearly document the origins of all the different types of placental mammals, but until then, the relationships between the orders must be inferred based largely on the characteristics of modern animals. For some time, biologists

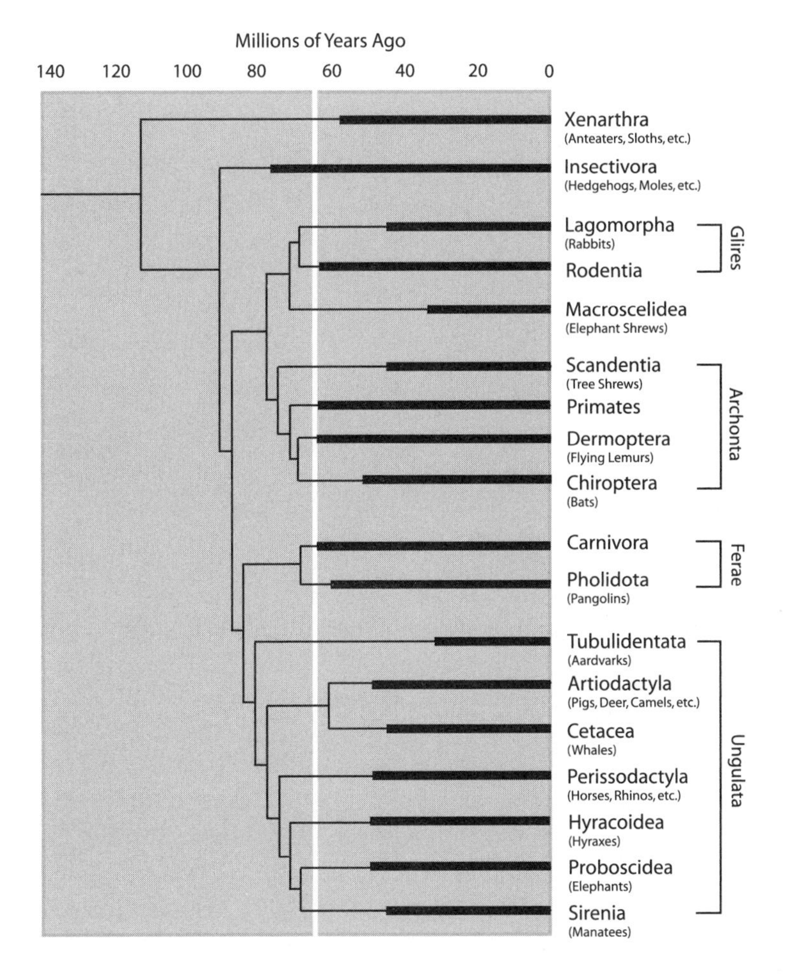

FIGURE 8.1 The fossil record and relationships of the placental mammal orders based on morphological (physical) characteristics. From Jeheskel Shoshani and Malcolm C. McKenna "Higher Taxonomic Relationships among extant Mammals Based on Morphology" *Molecular Phylogenetics and Evolution* 9 (1998): 572–584. The thick lines indicate how far back the fossil record of each group extends (breaks in these records and groups known only from fossils are not shown). The thin lines indicate the relationships between these orders based on comparing the morphological traits in these organisms. Note that the timing of the various forks in the tree is rather arbitrary, since it is very difficult to determine when the various groups branched from one another based only on morphological traits without sufficient fossil evidence. Indeed, based on the fact that most groups only date back to sixty-five million years ago (indicated by the vertical white line), when the dinosaurs died out, it has been suggested that most of these groups diverged around that time.

have attempted to devise mammalian family trees using detailed studies of rather subtle physical traits. The results of one such analysis can be seen in Figure 8.1. In this tree, the xenarthrans are the first order to branch off from the rest of the placental mammals, followed by most types of insectivores. The remaining placentals fall roughly into four major groups: Glires, which includes rodents and lagomorphs; Archonta, which includes bats and primates; Ferae, which includes carnivorans and pangolins; and Ungulata, which includes both types of hoofed animals, whales, and elephants. Within Ungulata, elephants, hyraxes, and manatees together form a group sometimes called Paenungulata.

Some of the relationships shown here seem intuitively plausible, such as the kinship between rodents and lagomorphs. Other connections, like those between bats and primates, are much less obvious. Unfortunately, most of these relationships are supported by a few characteristics that most would consider obscure, such as the shape of the contact between two bones in the skull. This means that, despite the fact that groups like Archonta and Ungulata have been assigned names, these categories might prove to be unreliable.

Data from DNA sequences has the potential to be a great help in this area. As we saw in the last chapter, differences between DNA sequences are easier to quantify than morphological characteristics, so they provide a promising way to recover relationships between diverse organisms. Furthermore, just as we could estimate when the ancestors of humans and chimps began to diverge using molecular data, it may even be possible to estimate when the various orders of mammals first emerged based on the number of differences between the relevant DNA sequences.

As was the case with the great apes, the raw data for this analysis is a collection of comparable DNA sequences, this time from a representative variety of living placental mammals—at least one from each order. Each of these DNA sequences is compared with all of the others in order to identify all the places where they differ, such as when ATGC in one sequence corresponds to ATTC in another. In the last chapter, we learned that the number of differences between any two sequences allows us to estimate how closely the two animals are related. However, remember that the previous analysis also assumed that the mutation accumulation rate in the relevant, noncoding, DNA sequences was roughly the same for all of the great apes. Now we are considering a far larger variety of animals, and there is evidence that some of these creatures have been accumulating mutations faster than others.

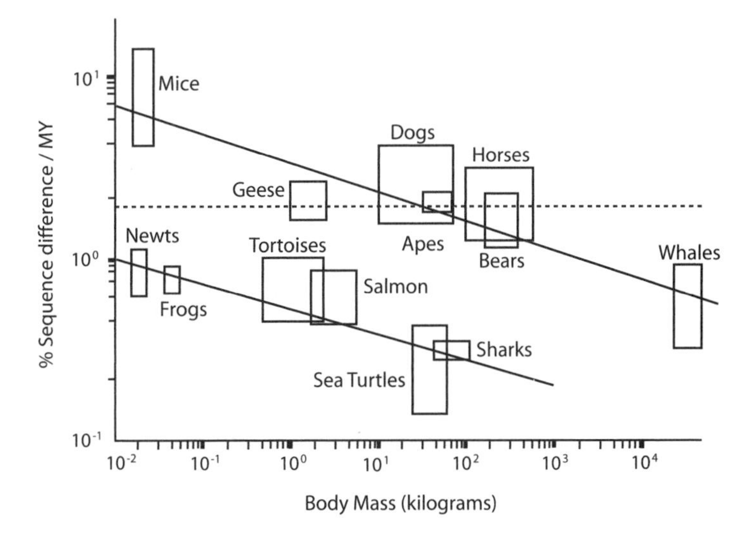

FIGURE 8.2 Different mutation accumulation rates in different animals. This figure is from A. P. Martin and S. R. Palumbi "Body Size, Metabolic Rate, Generation Time, and the Molecular Clock" *Proceedings of the National Academy of Sciences (USA)* 90 (1993): 4087. The mutation accumulation rate in mitochondrial DNA of different animals is plotted as a function of the body mass of the animal. Each box corresponds to a particular type of animal. The dashed line indicates where the boxes could lie if all animals accumulated mutations at the same rate. Note that larger animals have lower rates than smaller ones, and cold-blooded animals have lower rates than warm-blooded animals.

Figure 8.2 shows estimates of the mutation accumulation rates for a particular sequence in several different animals, based on a combination of molecular and fossil data. For example, in this particular sequence humans and chimps have different base pairs about 10% of the time. As we saw in the last chapter, the ancestors of humans and chimps began to split around six million years ago, so these animals have been accumulating mutations at a rate of approximately 2% per million years.[1] If mutations accumulated at this rate in all animals, then all of the animals would fall along a horizontal line on this graph. However, this is clearly not the case. In fact, the mutation accumulation rates range from as high as 5% per million years to as low as 0.2%. It is interesting to note that these variations are not random: large animals ap-

1. This is significantly faster than the 1% per five million years derived in the previous chapter, which indicates that different parts of the DNA sequence can accumulate mutations at different rates.

TABLE 8.3 Nucleotide differences between bears, dogs, llamas, and antelopes.

bear-dog	bear-llama	bear-antelope
36	45	50
	dog-llama	dog-antelope
	35	46
		llama-antelope
		38

parently tend to accumulate mutations more slowly than small animals, and cold-blooded creatures accumulate mutations more slowly than warm-blooded creatures. These and other trends observed in the mutation accumulation rates of various animals are still not fully understood.

To complicate matters further, most of the molecular data used to decipher the relationships between the placental mammals comes from coding regions of their DNA. In noncoding regions there is nothing stopping any mutation from being passed on to the next generation, so the accumulation rate is comparatively fast, up to several percent per million years. The fossil evidence indicates that most of the placental orders began to diverge over sixty-five million years ago, which means that every base pair in these sequences has probably acquired at least one mutation. This makes it almost impossible to match up the sequences between any two animals, and it also means that this DNA will contain almost no information about the common ancestry of these two organisms. We therefore need to use DNA sequences that accumulate mutations at a slower rate, which are found in genes. Remember that these sequences contain information for making proteins, so most mutations in these regions are unlikely to be passed on to future generations because they will adversely affect the organism's health. This reduces the mutation accumulation rate to an acceptable level, but it also complicates things by increasing the variability in the rates among these sequences.

Happily for us, we can still discover relationships and estimate divergence times even if mutations are accumulating at different rates in different animals. To illustrate this, we will use a particular sequence of about 300 nucleotides from a dog, a bear, a llama, and a Bongo antelope.[2] As with the apes discussed in the last chapter, we can compare these sequences and calculate the number of differences between each pair of animals (Table 8.3).

2. The DNA sequences themselves can be found on the PubMed database (www.pubmed.org), accession numbers AY011250, AY011249, AY011239, and AY011240.

If all of these creatures accumulated mutations at the same rate, then we would expect that the two animals with the smallest number of differences between them are the two animals that are most closely related to each other. In this case, the smallest number of differences is between the dog and the llama. This is surprising, because we would expect the dog to be most closely related to the bear, since they both belong to the order Carnivora. Similarly, we would expect the llama to be more closely related to antelope because they are both artiodactyls.

Even without knowing anything about the physical appearances of these animals, a closer look at the genetic data would demonstrate that that the dog and the llama are unlikely to be closely related. First, look at the top row of the above table. This says there are 36 differences between the bear and the dog, and 45 differences between the bear and the llama. If the dog and the llama really had the most recent common ancestor, then this means the ancestors of the llama acquired more mutations than the ancestors of the dog; the llama's lineage would then have accumulated mutations faster than the dog's. However, if we look at the last column of the same table, we find that there are 46 differences between the dog and the antelope, and only 38 differences between the llama and the antelope. This would indicate that it was the *dog's* lineage that was accumulating mutations at the faster rate, not the llama's lineage. The comparisons involving the bear and the antelope therefore directly contradict each other if the dog and the llama are assumed to have the most recent common ancestor.

Such a glaring inconsistency does not occur if we instead assume that the dog and the bear are more closely related to each other than either one is to the llama or the antelope. The llama has 45 differences from the bear and only 35 from the dog, indicating that the ancestors of the bear accumulated mutations more rapidly than those of the dog. Similarly, there are a larger number of differences between the antelope and the bear than between the antelope and the dog. This level of consistency indicates that the dog and bear do in fact share a close relationship. In a similar way, biologists can infer the correct relationships between animals even if the rate of mutations is not constant.

You may notice that in this particular sequence, the bear's lineage accumulated more mutations than the dog's, in contrast to the general trend shown in Figure 8.2. Furthermore, we get slightly different estimates of the mutation accumulation rate variations between dogs and bears depending on whether we compare these carnivores' sequences with the llama or the antelope. These residual discrepancies occur because a large fraction of the

300-odd nucleotides are different in the different sequences, so we cannot assume that each nucleotide has undergone only a single mutation. Such complications often arise in the analysis of real sequence data. However, with the appropriate mathematical tools to evaluate the quality and consistency of the data, biologists are still able to uncover the proper relationships between these organisms.

I should also point out that the above analysis is just one possible way to infer relationships from DNA sequences. Other methods, instead of relying on the total number of differences between organisms, look at each difference individually and assume that any individual mutation is unlikely to appear more than once. For example, if the bear and the dog have the sequence ATTG where the llama and the antelope have ATCG, it is more probable that the two carnivores are derived from a single ancestor with the sequence ATTG than they both acquired this same mutation independently. We can therefore deduce the relationships between the organisms by finding the phylogenetic tree that minimizes the number of coincidental mutations.

In the analysis of actual DNA data, computer programs find the most likely dendrogram. The algorithms that use the total number of differences between sequences are known as distance methods, while those which include data from individual mutations are known as maximum parsimony and maximum likelihood methods. These techniques all have their own advantages and disadvantages, and their relative usefulness in different situations is a subject of active debate.

Figure 8.3 shows the results of one such analysis that used the data from over 15,000 base pairs and more than 40 mammals. Some of the groups observed in Figure 8.1 also appear here: rodents and lagomorphs again form a well-defined group; and elephants, hyraxes, and manatees are once more found to be relatively closely related. Also, artiodactyls and whales turn out to be very closely related, which is consistent with the fossil evidence. In fact, these analyses tell us that some artiodactyls—like hippos—are more closely related to whales than they are to other artiodactyls, such as llamas and pigs. Consequently these two orders are here considered a single group called Cetartiodactyla.

At the same time, there are many relationships and groupings that were not seen in the earlier tree. In particular, the molecular data suggest that all placental mammals fall into one of four groups: Afrotheria, Xenarthra (consisting only of the order with that name), Euarchontoglires, and Laurasiatheria. All of the animals within any group are more closely related to each other than any of them is to an animal in another group. These four categories have appeared in

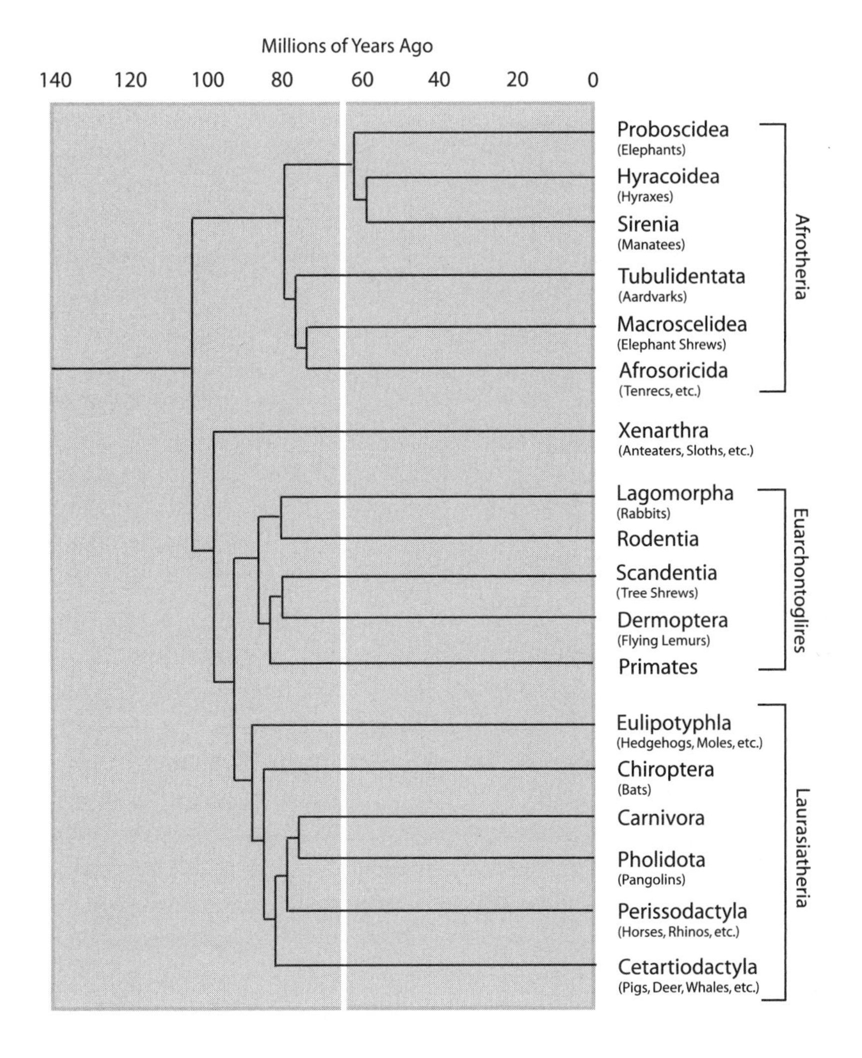

FIGURE 8.3 The relationships of placental mammals based on molecular data. From M. Springer et al. "Placental Mammal Diversification and the Cretaceous-Tertiary Boundary" *Proceedings of the National Academy of Sciences (USA)* 100 (2003): 1056–1061. The relationships derived through this method are not exactly the same as those shown in Figure 8.1. In this figure, the ages of the various groups are based on molecular data, which indicate that the orders diverged well before sixty-five million years, when the last of the dinosaurs died out. Note that Cetacea and Artiodactyla are fused into a new order; this is because whales are more closely related to certain artiodactyls, like hippos, than hippos are to other artiodactyls, like pigs. Eulipotyphla and Afrosoricida are two groups of Insectivore families: Enriceidea + Talpidae + Soricidea + Solenodontidea = Eulipotyphla; Tenrecidae + Chrysocholoridae = Afrosoricida.

a number of molecular studies, and they are supported by the fact that members of each of these groups share certain mutations—such as a deletion of a particular stretch of base pairs—that are very unlikely to occur more than once.

There is still some uncertainty about the relationships between these groups, but the tree presented here is one of the better-supported options at the moment. In this case, the first group to branch off from the others is Afrotheria, which includes elephants, hyraxes, manatees, and aardvarks. It also contains the elephant shrews and two families of insectivores, the tenrecs and the golden moles. As its name implies, most of these animals are found in Africa.

Next is Xenarthra, the order that includes sloths, anteaters, and armadillos. Today, most of these animals are found in South America.

Finally, Euarchontoglires and Laurasiatheria diverge. Euarchontoglires contains rodents, rabbits, primates, flying lemurs, and tree shrews. In other words, it contains all members of the previously mentioned groups Glires and Archonta except for the bats.

Laurasiatheria is the most diverse group. It contains bats, carnivorans, pangolins, whales, and both artiodactyls and perissodactyls. It also includes the insectivore families Erinaceidae, Talpidae, and Soricidae: the hedgehogs, shrews, and moles.

On the surface, this set of relationships may not look any more or less plausible than the relationships based on physical traits given in Figure 8.1. However, this fourfold division of placental mammals shows some intriguing geographical patterns. Afrotherians are from Africa, xenarthrans are from South America, and the fossil record suggests that euarchontoglires and laurasiatherians—which are found throughout the world today—originated in the northern continents. This possible connection between mammalian relationships and geography has caught the interest of many evolutionary biologists, because it potentially provides important clues about the early history of placental mammals.

SECTION 8.3: FINDING TIME WITH BAYESIAN STATISTICS

To explore further the possible relationships between geography and mammalian evolution, we need to know when the various groups began to diverge from each other. Fossils of the earliest known representatives of most orders date back to around sixty-five million years ago, when the last of the dinosaurs died out. Of course, there were mammals around before this time, but the relationships between these early creatures and modern orders is not clear.

In fact, some paleontologists argue that most or all of the orders of placental mammals really did arise at about the same time sixty-five million years ago, when mammals began to fill the many vacant ecological niches left by the extinction of the great dinosaurs, etc. This scenario is now being challenged by researchers who have attempted to estimate when the various branches of the mammalian family tree occurred using gene sequence data.

At first, it may appear that using genetic data to measure the passage of time would be impossible in this case. After all, we have seen that different mammals accumulate mutations at very different rates. However, careful analysis of the sequence data often reveals the variations in the mutation accumulation rates of different lineages. In the simple example above, the number of differences between the animals was sufficient to establish that—in one particular sequence—the ancestors of bears accumulated mutations faster than the ancestors of dogs. The DNA sequences therefore contain information that biologists can use to correct for variations in the mutation accumulation rate.

Even with this information, extracting the timing information from the sequence data is not a simple matter. If we knew the mutation accumulation rate for each branch, it would be fairly easy to calculate the number of differences that should exist between the animals today. However, working backwards and using the observed differences to infer the relevant rates is a far more challenging task. There are a number of different combinations of rates that *could* have produced the observed differences between sequences, and we want to figure out which of these possibilities is most likely to be what *actually* occurred. To accomplish this, some molecular biologists have turned to an approach known as likelihood analysis or Bayesian statistics.

To demonstrate how Bayesian statistics works, let us consider the flipping of a coin. If we were told a coin was flipped 10 times, and we wanted to know the chances of getting 5 heads, then we have a typical statistics problem. All one needs to do to solve it is to write down all the possible results of 10 coin flips and then calculate what fraction of these give 5 heads. If you remember your combinatorics, you can avoid the tedious step of writing down all the possible results, but either way, you get the correct answer. Similarly, it is a trivial matter to get the chances of getting 5 heads if you flip a coin 6 times, 100 times, or 1,000 times.

However, what if we knew that the coin came up heads 5 times, and we wanted to estimate how many times the coin was flipped? You would probably guess that the coin was flipped about 10 times, since we know it should have come up heads in about half of the flips. However, most statistical techniques do not allow us to calculate the chances that the coin was flipped 10 times, as opposed to 8 times or 20 times. Bayesian statistics can help us solve this problem at the cost of a single assumption, known as the prior. In this case, we may assume that the coin could have been flipped any number of times. In the absence of any other information about the situation, this prior assumption seems perfectly reasonable.

With this assumed prior, we can now calculate the probability that the coin was flipped a given number of times *and* that the coin came up heads 5 times. Boiled down to an equation, this number is the chance that the coin was flipped the specified number of times multiplied by the chance of getting 5 heads with that number of flips. Operating under the assumption that the coin could have been flipped any number of times, the first probability is a constant with the number of coin flips. On the other hand, the chance of getting 5 heads is not the same for all total flips. For instance, there is no way to get 5 heads when we flip the coin less than 5 times. If the coin is flipped 5 times, it is improbable that it will come up heads every time. On the other hand, if the coin was flipped 100 times, it is even more improbable that it would come up heads just 5 times. It is only when the coin is flipped around 10 times that we are reasonably likely to have the coin come up heads 5 times. Therefore, even though we have assumed the coin *could have been* flipped any number of times, the fact that heads came up 5 times indicates that the coin most likely was *actually* flipped around 10 times. If we like, we can even determine the relative likelihood of any number of coin flips using this approach.

Even though these calculations rely on the assumption of a prior probability in the number of coin flips, there is a great deal of latitude in the types of priors we can assume. Above we assumed that all numbers of coin flips were equally likely, but we could also have assumed that the coin could only have been flipped as many as 30 times. Since the chance of getting just 5 heads with more than 30 flips is so low, this assumption does not strongly affect the result; 10 coin flips is still the most likely answer. In fact, the only prior assumptions that can alter the calculation are those that excessively limit the number of coin flips, or demand that the number of coin flips must be much larger than 10. Without additional information, there is no basis for using such odd priors. This method then provides reasonably accurate results, allowing scientists to explore problems where it is more straightforward to calculate the probability of getting some result of a given process than it is to compute something about the processes that led to that result.

Bayesian statistics can even be applied to the question of mammalian evolution. In this case, the assumed prior could go something like this: each branch of the tree shown in Figure 8.3 could have appeared at any time, so long as the order of the branches is preserved, and the mutation accumulation rate in any branch could be any value. The relative likelihood that a given set of branching times and accumulation rates are the actual values can then be calculated as the probability that those times and rates would have produced the observed differences in the DNA sequences.[3] Of course, there are too many rates and times to calculate the probabilities for them all, so biologists use algorithms to search for the most likely values, making some additional assumptions that the rates do not vary wildly from one branch to the next.

One additional benefit of this approach is that it enables data from the fossil record to be incorporated directly into the calculation of various timescales. For example, fossils of early whales suggest that cetaceans began to diverge from terrestrial artiodactyls over fifty million years ago. Phylogeneticists can account for this in the Bayesian analysis by imposing the prior that the whale's branch cannot diverge from the hippo's branch (or whatever is assumed to be the whale's closest relative) more recently than fifty million years ago. This information not only helps constrain the variations in the mutation accumulation rates, it also provides the reference points needed to obtain ages in actual years.

This method of obtaining the ages of different lineages with variable mutation rates is new, and it will be some time before biologists can ascertain if it is sufficiently reliable. Even if the DNA sequence data yields the correct pattern of branches in the mammalian family tree, the additional assumptions required to compute the dates could lead to incorrect results. A number of researchers are currently working to test and refine this dating method. In the meantime, it is still interesting to consider the results of one of these analyses—illustrated in Figure 8.3—which indicates that most of the orders of placental mammals had already established themselves before sixty-five million years ago. In other words, these groups had arisen before the giant dinosaurs died out. This does not mean that we had mammals that looked like today's horses or monkeys frolicking around with *Tyrannosaurus rex*, but it does imply that different orders arose from separate stocks of Mesozoic mammals.

Another intriguing result of these analyses is that the ancestors of Afrotherians—a group now largely confined to Africa—branch off from the ancestors of other modern placental mammals roughly 105 million years ago. This division occurs at a pivotal time in the geography of the Southern Hemisphere. South America and Africa were originally joined together as part of a larger

3. It is even possible to include different patterns of branches into this analysis.

continent called Gondwana. However, roughly 100 million years ago South America and Africa were pulled apart by continental drift, and became separated by a young Southern Atlantic Ocean (see Figure 8.4). If there was a population of primitive placental mammals in Gondwana around 110 million years ago, this geological rupture would have produced two more or less isolated groups of mammals, one in Africa and another in South America. The former would then become the ancestors of afrotherians, and the latter would then have to be the ancestors of the xenarthrans, euarchontoglires, and laurasiatherians. We can even imagine that a few million years later some of the South American mammals found their way into the Northern Hemisphere, perhaps along a chain of islands between North and South America that would eventually become Cuba, Hispaniola, and Puerto Rico. The animals that made this intrepid journey would be the ancestors of the euarchontoglires and the laurasiatherians, while those they left behind would give rise to the xenarthrans.[4]

The question now is whether this picture of globe-trotting Mesozoic mammals is consistent with the fossil evidence. As we mentioned above, the earliest animals with traits diagnostic of particular orders appear only around sixty-five million years ago. However, perhaps the earlier mammals simply had not yet evolved these distinctive traits, so their relationships with modern mammals has been obscured. To explore this possibility, paleontologists are examining those fossils from the age of the dinosaurs that appear to belong to creatures closely related to modern placental mammals. Of course, such ancient remains do not provide clear evidence that these animals had a truly placental mode of reproduction. However, the teeth and bones of these creatures contain features that set them apart from marsupials, monotremes, and other types of mammals and indicate that they share an ancestor with modern placentals. Since it would be inappropriate to assume the earliest members of this group already had a fully functional placenta, the more neutral term "eutherians" is used to refer to the members of this entire branch of the mammalian family tree, which includes modern placental mammals.[5]

4. Some DNA studies suggest that the xenarthrans were the first group to diverge from other mammals, followed by the afrotherians. In this case, the ancestors of the euarchontoglires and laurasiatherians would have moved north from Africa instead of South America.

5. Some paleontologists use the word "Eutheria" to refer to all animals more closely related to placental mammals than to modern marsupials, and the word "Placentalia" to refer to all animals descended from the last common ancestor of all modern placental mammals. In this situation, all modern placental mammals belong to both Placentalia and Eutheria, but some fossil animals could be eutherians and not placentalians.

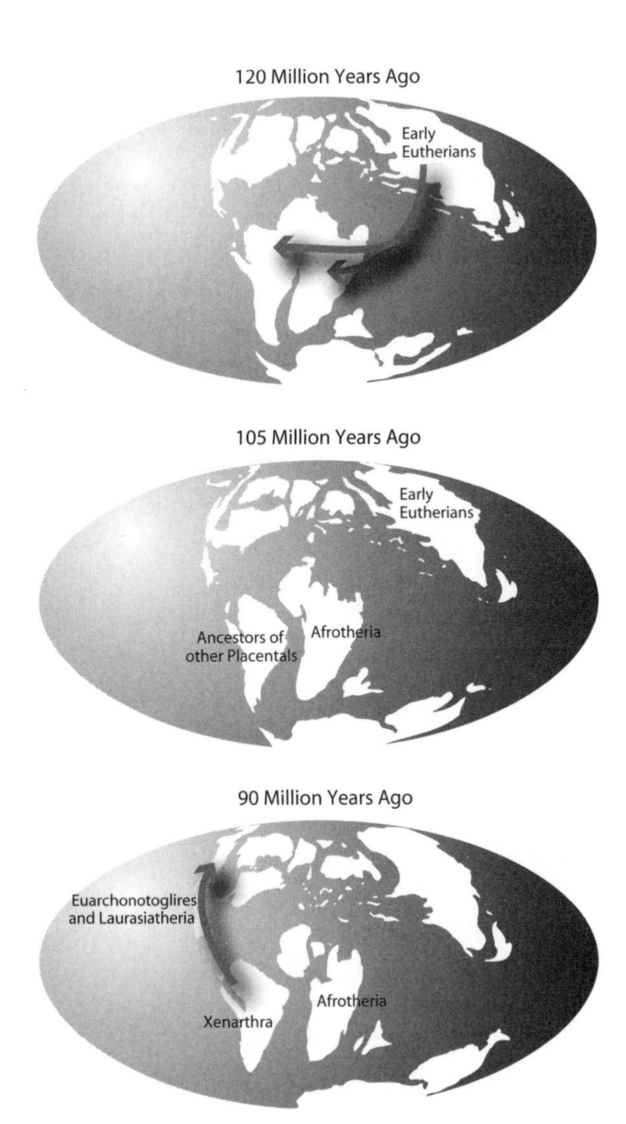

FIGURE 8.4 One possible scenario for the movements of early placental mammals during the end of the age of dinosaurs. These maps show the continents at different times in the past, together with a possible distribution of eutherian mammals based on evidence from the DNA sequence data. First, some early eutherian mammals find their way into the southern hemisphere. Around 105 million years ago, Africa and South America separate, isolating the ancestors of Afrotherian mammals from other groups. 5–10 million years later, some of the animals in South America migrate into the northern continents, leaving behind the ancestors of Xenarthra. The mammals that arrived in the north eventually became the Euarchontoglires and Laurasiatherians.

The earliest known eutherian comes from China. It was described in 2002, based on a well-preserved specimen consisting of a nearly complete skeleton and, remarkably, imprints of the animals' fur. This creature, named *Eomaia scansoria*, was about twenty centimeters long from snout to tail and can be recognized as a eutherian based on the structure of its anklebones. It lived around 125 million years ago, which means eutherians did indeed exist as a distinct group of mammals over 20 million years before the ancestors of modern placentals are supposed to have begun to diverge from each other. In this respect, at least, the fossil evidence is consistent with the molecular data. However, *Eomaia* does not have any traits that indicate that it is directly related to any particular group of modern placental mammals, so this fossil cannot tell us whether groups like Afrotheria or Xenarthra really did appear during the Mesozoic.

There are some fossils that may provide evidence for the antiquity of some orders of modern eutherians. These remains belong to two types of creatures—called zalambalestids and zhelestids—that lived primarily in Asia 75 to 90 million years ago. Recently some paleontologists have argued, based on a detailed study of their teeth, that the zalambalestids are closely related to the ancestors of rodents and rabbits, and that the zhelestids are closely related to hoofed animals. However, these results are still very controversial because of the limited number of modern mammals included in the analysis. Efforts are currently under way to perform more complete analyses of the relationships between early and modern eutherian mammals, some even incorporating information from both DNA sequences and physical traits. Combining such heterogeneous data sets presents many difficulties, but hopefully it will lead to a robust determination of the relationships between modern placental mammals and their Mesozoic ancestors. Until then—or until new fossils clarify the relationships between these early mammals—it will remain an open question whether or not any of these ancient eutherians were ancestors of specific groups of modern placental mammals.

A much greater challenge to the above molecule-based picture of the spread of early placental mammals has to do with *where* the early eutherian fossils have been found thus far. Almost all of the fossils of eutherian mammals that are more than ninety million years old—including *Eomaia*—come from Asia. Slightly more recent deposits in Europe and North America also contain eutherian mammals. Very few early eutherians have been found in either South America or Africa. Even the earliest known relatives of afrotherians are found in the northern continents. Many paleontologists are therefore very skeptical of the idea that all modern placental mammals trace their origins back through the Southern Hemisphere. Of course, the fossil deposits in the southern continents are in general not as well explored as their northern counterparts, so future discoveries in Africa or South America may eventually confirm the story of early placental mammals presented in Figure 8.4.

Alternatively, some researchers have argued that instead of the ancestors of the euarchontoglires and laurasiatherians going north, the afrotherians and xenarthrans may have come south to Africa and South America from the northern continents. Such an idea would be more consistent with the available fossil evidence. However, for this scenario to be consistent with the relationships inferred from the molecular data, all modern euarchontoglires and laurasiatherians must be descended from a stock of early placental mammals remaining in the Northern Hemisphere that were more closely related to the mammals that arrived in South America than those that reached Africa (or vice-versa, according to some analyses). It is not at all obvious why this would be the case.

Obviously, our understanding of the origins of placental mammals is still far from complete. Even so, the new molecular data has sparked a renewed interest in the subject and has even suggested promising new areas for future research. For example, some biologists are particularly intrigued by the large number of examples of convergence and parallelism that appear between the four major groups in the above phylogenetic tree: anteaters in Xenarthra, aardvarks in Afrotheria, and pangolins in Laurasiatheria all have similar adaptations to eating ants and termites, while the moles in Laurasiatheria, the mole rats in Euarchontoglires, and the golden moles in Afrotheria are all well adapted for living underground, just to give two examples. If the ancestors of these different groups of mammals were established in different corners of the world at the time when the last of the giant dinosaurs died out, we can imagine that generally shrew-like mammals in different regions of the world all began diversifying and evolving to fill a wide variety of ecological niches. This means that different animals at different places were adapting to exploit the same resources at roughly the same time, and they eventually acquired similar advantageous traits. Comparing these animals and their lineages would then provide important clues about how such characteristics evolved. Efforts to unravel the early history of the eutherians will therefore certainly yield many fresh insights into the processes that have shaped the history of life.

Future fossil discoveries and more refined analytical techniques should eventually allow biologists to produce a detailed account—consistent with both the morphological and the molecular data—of the distribution, relationships, and characteristics of the placental mammals who lived before the arrival of that fateful meteorite sixty-five million years ago. Whatever role that impact played in the fall of the dinosaurs and the rise of the mammals, it certainly opened a new chapter in the story of life on earth. It also marks the end of the line for a object that may have been flying around space for billions of years. As we will see in the next chapter, meteorites are not just agents of destruction; they also often preserve an invaluable record of the early history of the solar system.

SECTION 8.4: FURTHER READING

For a nice popular summary of today's mammals, see E. Gould and G. McKay *Encyclopedia of Mammals* (Academic Press, 1998). For more technical details, try G. A. Feldhammer et al. *Mammalogy* (McGraw-Hill, 1999). For the fossil evidence for the evolution of mammals, see E. H. Colbert et al. *Colbert's Evolution of the Vertebrates* (Wiley, 2001), Robert Carroll *Vertebrate Paleontology and Evolution* (W. H. Freeman, 1988), and Joel Cracraft and M. J. Donoghue *Assembling the Tree of Life* (Oxford University Press, 2004).

For something more specifically about early mammals, see Z. Kielan-Jaworowska et al. *Mammals from the Age of Dinosaurs* (Columbia University Press, 2004) and K. A. Rose and J. D. Archibald *The Rise of Placental Mammals* (John Hopkins University Press, 2005).

For details on the newly discovered whale ancestors, see J. G. M. Thewissen et al. "Skeletons of Terrestrial Cetaceans and the Relationship of Whales to Artiodactyls" *Nature* 413 (2001): 277–281. For details on the discovery of *Eomaia*, see Q. Ji et al. "The Earliest Eutherian Mammal" *Nature* 416 (2002): 816–822.

Good websites for exploring the relationships between organisms are tolweb.org and palaeos.com; many useful articles can also be found at www.pubmed.org and www.plos.org.

For details of various methods for extracting phylogenetic trees from molecular data, see Wen-Hsiung Li *Molecular Evolution* (Sinauer, 1997) and M. Nei and S. Kumar *Molecular Evolution and Phylogenetics* (Oxford University Press, 2000).

For the latest work on dating the divergences of eutherian mammals with molecular techniques, see M. Springer et al. "Placental Mammal Diversification and the Cretaceous-Tertiary Boundary" *Proceedings of the National Academy of Sciences (USA)* 100 (2003): 1056–1061 and M. Hasegawa et al. "Times Scale of Eutherian Evolution without Assuming a Constant Rate of Molecular Evolution" *Genes and Genetic Systematics* 78 (2003): 267–283.

A recent article that presents a slightly different pattern of mammalian evolution is J. O. Kriegs et al. "Retroposed Elements as Archives for the Evolutionary History of Placental Mammals" by *PLoS Biology* 4 (2006): 0537. Some good review articles on the ongoing efforts to interpret this new data are M. Springer et al. "Molecules Consolidate the Placental Mammal Tree" *Trends in Ecology and Evolution* 19 (2004): 430–438, J. D. Archibald "Timing and Biogeography of the Eutherian Radiation: Fossils and Molecules Compared" *Molecular Phylogenetics and Evolution* 28 (2003): 350–359, and J. P. Hunter and C. M. Janis "Spiny Norman in the Garden of Eden? Dispersal and Early Biogeography of Placentalia" *Journal of Mammal Evolution* 13 (2006): 89–123.

Skeptical takes on the molecular data can also be found Cracraft and Donoghue *Assembling the Tree of Life* and Rose and Archibald *Rise of the Placental Mammals*, referred to above. An interesting article pointing out some fundamental limitations to the precision of genetic methods for measuring age is T. Britton "Estimating Divergence Times in Phylogenetic Trees without a Molecular Clock" *Systematic Biology* 54 (2005): 500–507.

Finally, another recent effort to use DNA data to explore when different stocks of mammals may have appeared around the end of the age of dinosaurs can be found in: O. R. P. Bininda-Emonds et al. "The Delayed Rise of Present-Day Mammals" *Nature* 446 (2007): 507–512.

CHAPTER NINE

Meteorites and the Age of the Solar System

Shortly before midnight on March 26, 2003, people living in the suburbs south of Chicago had some unexpected visitors. Earlier that evening a rock from outer space with a mass of at least 900 kilograms entered our atmosphere. As it got closer to earth, it slowed down, heated up, and eventually broke into many pieces. The smaller fragments vaporized in the atmosphere, but there were still hundreds of pebble- and cobble-sized meteorites left to rain down on sleeping neighborhoods, crashing into sidewalks, houses, and cars. The biggest objects—with masses up to 5 kilograms—fell in a village called Park Forest, so this material became known collectively as the Park Forest meteorite.

Of course, this was not the first time rocks from space have crash-landed on earth. Much larger objects—like the one that may have played a role in the demise of the giant dinosaurs—have collided with our planet over the last few billion years, and the possibility of a future large impact has both spurned efforts to catalog and monitor potentially hazardous objects and inspired many science fiction movies. But while the arrival of a meteorite can be quite a dramatic event, the object itself is also a precious resource for solar system research.

Except for the moon rocks brought back by the Apollo and Luna missions and some smaller cometary grains returned by the Stardust spacecraft, meteorites are the only nonmicroscopic samples from outer space available for us to study here on earth. Furthermore, many meteorites are like time capsules, containing various little bits of rocky debris from an era when the planets had not yet fully formed. These tiny relics display a variety of chemical compositions and physical characteristics, indicating that they formed under different conditions and at different times. The chemistry, mineralogy, and even age of these objects therefore provide important clues about the origins and early history of our solar system.

SECTION 9.1: ROCKS FROM THE SKY

In all, there are over fifty different types of meteorites. The most spectacular are the so-called iron meteorites, which are composed largely of iron and nickel alloys, and the stony-irons, which contain mixtures of metal and rock. When cut open and polished or etched, these objects can reveal intricate hexagonal patterns or a beautiful array of greenish crystals suspended in a metallic background. However, only about 5% of meteorites contain large amounts of metal, and the vast majority are essentially rocks made up mainly of silicate minerals. These stony meteorites are divided into two broad groups based on whether their internal structure includes chondrules, small spheres of rock roughly a millimeter across. Stony meteorites possessing these rocky grains are known as chondrites, and those without them are called achondrites.

Chondrites are the most common class of meteorites, making up almost 90% of the known finds. They also have particularly interesting chemical compositions. The mixture of elements found in certain classes of chondrites is remarkably similar to that of the sun. In other words, these rocks and the sun have roughly the same relative amounts of silicon, iron, magnesium, sodium, nickel, phosphorus, and so on. The only major difference between the composition of the chondrites and the makeup of the sun is that the sun has higher concentrations of hydrogen, helium, carbon, oxygen, nitrogen, neon, and argon.[1] These elements are all either light or chemically unreactive, so they can easily escape from objects that lack a strong gravitational field like the sun. Cosmochemists (the people who study these things) therefore suspect that both chondrites and the sun formed from the same materials. This implies that chondrites are part of our solar system and not interstellar interlopers.

Their composition also suggests that chondrites are relics from a very early stage in the history of the solar system, when the planets were just beginning to congeal out of a disk of dust and gas swirling around the young sun. Our entire solar system probably arose from an enormous cloud of dust and gas that collapsed under its own gravity. If this cloud was rotating before it started to collapse, this circulating motion naturally favors the formation of a central star surrounded by a flattened disk. Indeed, recent observations confirm that just such a disk surrounds many young stars. This disk—which was composed of essentially the same stuff as the star at its center—then supplied the raw material for

1. Chondrites also tend to have higher lithium fractions than the sun, likely because the nuclear reactions that power the sun tend to consume lithium.

the planets, moons, asteroids, and comets found in the solar system today. However, while both chondrites and planets were probably formed from the same material, rocks found on the earth and other planets do not normally have chondritic compositions, due to a variety of processes that altered the distribution of elements in large objects. For example, much of the earth's iron is now located in the depths of its core, while its crust is enriched in elements like silicon. This differentiation reflects a global redistribution of material that likely occurred early in earth's history, when heating from radioactive decay and collisions with other objects left the planet in a partially molten state, allowing elements with different masses and chemical properties to segregate. Similar processes appear to have affected all of the planets, most of the moons, and even certain asteroids. Since chondrites seem to have been able to avoid these phenomena, these extremely ancient objects can provide a window onto the origins of the solar system.

SECTION 9.2: RADIOMETRIC DATING FOR METEORITES

Like the prehistoric organic matter and the volcanic rocks described in previous chapters, scientists deduce the age of meteorites from the decay of unstable nuclei. However, the radiometric techniques used with meteorites are not exactly the same as the carbon-14 and potassium-argon dating methods discussed earlier. Meteorite dating both uses different unstable isotopes and employs a distinct—and very clever—technique to infer how many unstable nuclei were originally in the material.

Dating meteorites poses particular challenges because there is no nonradiometric data set available that can be used to document the original isotopic composition of meteorites, which means the ages of meteorites cannot be calibrated like carbon-14 dates. Also, unlike most rocks on earth, meteorites have been involved in collisions before they were sent hurtling towards us. Such events could have been violent enough to jostle loose some or all of the noble gases from the rock. Because of this, techniques like potassium-argon dating do not necessarily yield a reliable estimate of when the meteorite itself formed.

If we wish to estimate the age of a meteorite, we need to use radioactive isotopes that decay into elements that cannot easily escape from the rock during its tumultuous life. One such isotope is rubidium-87, an unstable variant of the rather obscure element rubidium. This atom undergoes beta decay and becomes a stable variant of strontium—strontium-87, to be exact. The half-life for this decay is about 50 billion years, so we can use this nucleus to measure the age of very old rocks like meteorites. Unlike the argon in the potassium-argon system, which is usually chemically inert, both rubidium and strontium are chemically reactive elements. All of the strontium-87 produced by the decay of rubidium-87 therefore remains in the rock even if it is exposed to severe collisions.

However, the chemical properties of strontium-87 also complicate efforts to extract an age from the rock. Imagine we find a rock that contains 300 milligrams of rubidium-87 and 300 milligrams of strontium-87. We then know it always had a total of 600 milligrams of some combination of rubidium-87 and strontium-87. However, unlike argon-40, strontium-87 is chemically reactive, so there is the possibility that some amount of this isotope entered the meteorite as it formed. With just these two numbers, it is impossible to tell how much of the strontium-87 was there from the beginning and how much is the product of nuclear decay. Therefore, we cannot say whether the rock initially had 600, 500, or 350 milligrams of rubidium-87. We need more information if we want to use the rubidium and strontium content of the rock to estimate its age.

Fortunately, rocks, meteorites, and even chondrules are not homogeneous objects, but are made up of a variety of different minerals. By measuring the rubidium and strontium contents of the different components of the rock, we can extract sufficient information to obtain a date. For example, say we broke a meteorite into two pieces with different mineral compositions, we could then measure the amount of rubidium-87 and strontium-87 in both of these pieces. Also, for reasons that will become clear in a minute, we measure the amount of another isotope of strontium, strontium-86, which is not produced by nuclear decay of another element. Hypothetically, say that the compositions of these two pieces of the meteorite are:

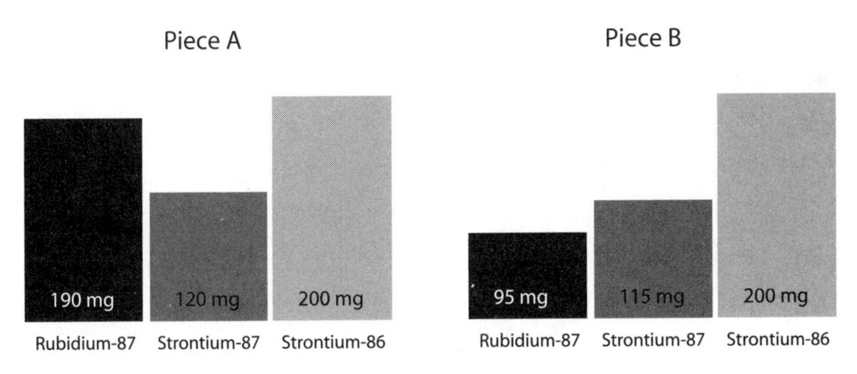

FIGURE 9.1

Piece A and piece B happen to have the same amount of strontium-86, and different amounts of rubidium-87. Since rubidium and strontium have different chemical properties, it is not surprising that different parts of the meteorite would have different mixtures of these elements. However, the sample with more rubidium-87 also has more strontium-87. This suggests that some of the strontium-87 is the result of nuclear decay and that the rock formed some time ago.

Imagine that at some point in time, we had a blob of molten rock floating in space. As this material cooled, various minerals within it began to solidify. Due to differences in their chemistry, these minerals incorporated different amounts of rubidium and strontium into their crystal structures. By contrast, strontium-86 and strontium-87 are isotopes of the same element and so any mineral should have absorbed both of them at the same rate. This means that right after the meteorite formed, both of the samples should have contained the same amount of not only strontium-86 but also strontium-87. As time goes on, a portion of the rubidium-87 decays, producing additional strontium-87. Piece A contains more rubidium-87 than piece B, so piece A has been able to accumulate more strontium-87 than piece B.

Since we know the current amounts of rubidium and strontium in the two pieces and the half-life of rubidium-87, we can estimate how much rubidium-87 and strontium-87 was in the two pieces at any time in the past. It takes approximately 3.5 billion years for 5% of the rubidium-87 to convert to strontium-87. Therefore, 3.5 billion years ago the samples contained 5% more rubidium-87 than they do now. Piece B would therefore have 100 milligrams of rubidium-87 instead of 95, and piece A would have 200 milligrams instead of 190. The compositions of the two pieces of the rock 3.5 billion years ago would therefore be:

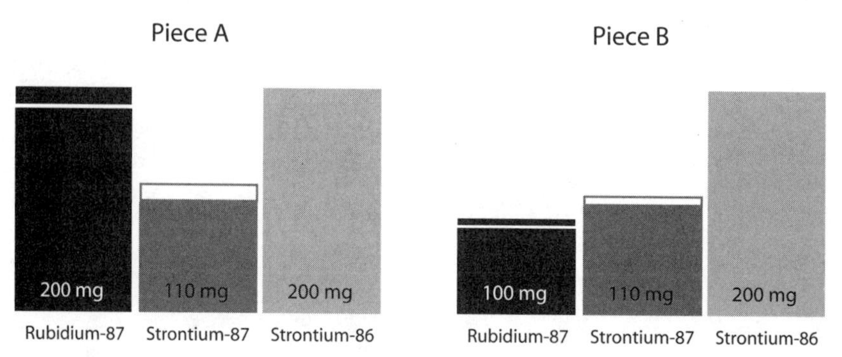

FIGURE 9.2

According to these calculations, the two parts of the meteorite would have the same amounts of strontium-87 (110 mg) relative to strontium-86 (200 mg) 3.5 billion years ago. This congruence is what we would expect if the minerals had just solidified, so these data indicate that the rock could have formed 3.5 billion years ago. Doing similar calculations for other decay percentages

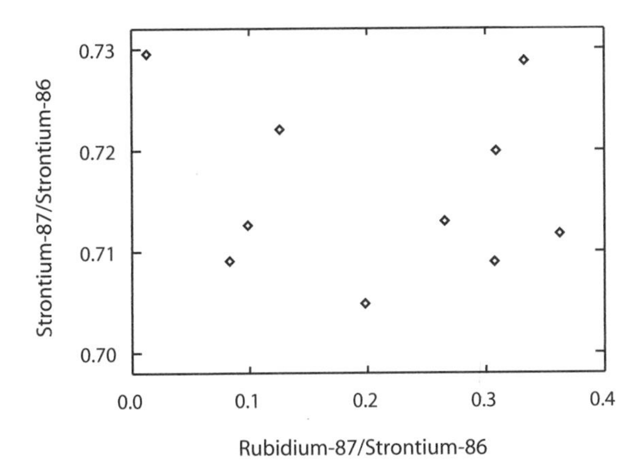

FIGURE 9.3 The isochron plot of a rock cobbled together from minerals formed at different times and different environments.

demonstrates that this is the *only* time when the two samples would have the same mix of strontium-86 and strontium-87. If both of these pieces coalesced at the same time in the same environment, they—and the meteorite—could only be 3.5 billion years old.

The above analysis is a very basic form of isochron dating, a technique that is used extensively in the study of meteorites and other ancient rocks. In this simple example, we had to assume that both parts of the rock formed at the same time and in the same environment. In practice, there is usually sufficient information in the rocks to determine whether the various minerals formed at once or not, so it is actually possible to verify this assumption with data from the rock itself. To realize the full potential of this dating method, cosmochemists must measure the rubidium-87, strontium-87, and strontium-86 contents of many different minerals in the material. With these data, we can calculate how many grams of rubidium-87 or strontium-87 there are per gram of strontium-86 in each sample, and then make a graph of the rubidium-87 content of the minerals versus their strontium-87 content. These sorts of graphs, commonly known as isochron plots, not only document the age of the rock but also indicate whether that date is reliable.

For example, imagine we had a meteorite that was cobbled together out of minerals that formed at different times and in different environments. In this case, the strontium-87, strontium-86, and rubidium-87 contents of these minerals would not have any obvious relationship to one another and the isochron plot would look like Figure 9.3. Each point on this plot represents the com-

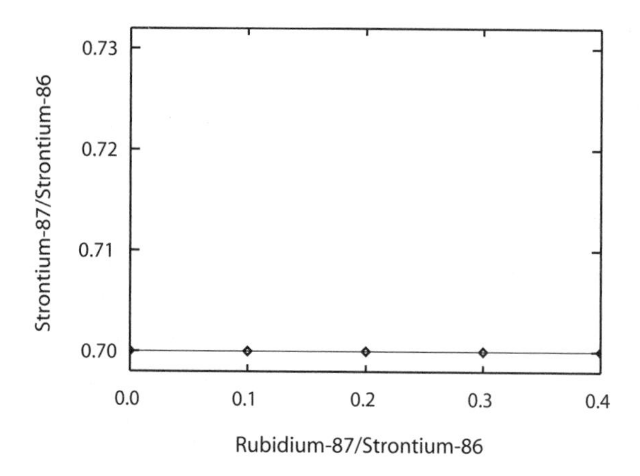

FIGURE 9.4 The isochron plot of a collection of minerals that recently formed from a single source.

position of a single mineral in this hypothetical meteorite. Its position along the vertical axis gives the ratio of strontium-87 to strontium-86, and its position along the horizontal axis indicates the ratio of rubidium-87 to strontium-86. Notice that for this random collection of minerals, the points are scattered haphazardly, graphically demonstrating the lack of any relationship between the strontium-87 and the rubidium-87 content of the minerals.

Next, imagine that we had a collection of minerals that had recently formed from a single source. If this source contained seven grams of strontium-87 for every ten grams of strontium-86, then all of the minerals would also have this same mix of strontium isotopes. The isochron diagram of such a collection would look like Figure 9.4. Since all of the minerals have the same strontium-87/strontium-86 ratio, the points fall along a horizontal line. The distribution of points on an isochron plot therefore depends on whether the minerals have a common source or not. This distinction remains even as the rock ages.

Now suppose we could let this rock sit for 5 billion years and then remeasure the rubidium and strontium contents of the various minerals. During this time, around 7% of the rubidium-87 will have decayed into strontium-87. This means the sample that originally had 0.3 grams of rubidium-87 per gram of strontium-86 will now have only 0.28 grams of rubidium-87 and 0.02 additional grams of strontium-87 per gram of strontium-86. By the same token, the mineral with an original rubidium-87 content of 0.4 will today have a rubidium-87 content of about 0.37 and almost an additional 0.03 in its strontium-87 content. If we redraw the isochron plot we get something like Figure 9.5 (the light gray dots trace how the composition of each mineral has changed over the past 5 billion years). The points still fall along a line, but it is no longer horizontal because minerals that contained more rubidium-87 have a larger amount of strontium-87 generated through radioactive decay. Note that the line still intercepts the *y*-axis at 0.7, which is the original strontium-87/strontium-86 ratio for the rock. To understand why this happens, realize that the *y*-axis of this plot corresponds to a mineral with no rubidium-87 at all, and the strontium-87 content of such a mineral never changes. The *y*-intercept of this line therefore provides a measure of the original strontium-87 content of all of the minerals. With this number in hand, we can compute how much strontium-87 in any mineral is due to the decay of rubidium-87 and determine the age of the meteorite. Alternatively, we can find the age of the rock by simply measuring the slope of the line, which steadily increases as the rock ages.

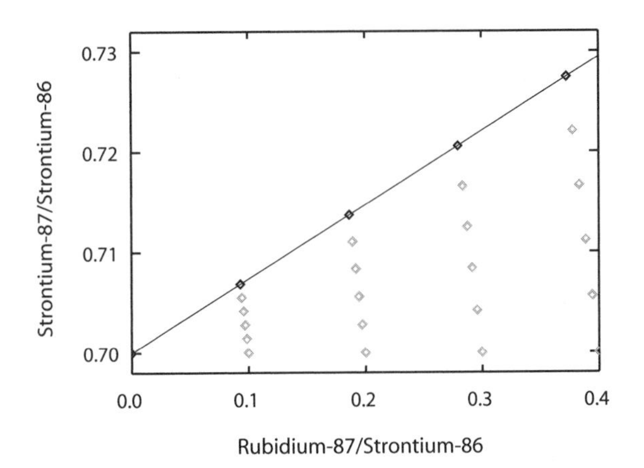

FIGURE 9.5 The isochron plot of a rock formed from a single source five billion years ago. (The light gray dots trace how the composition of each mineral has changed over the past five billion years.)

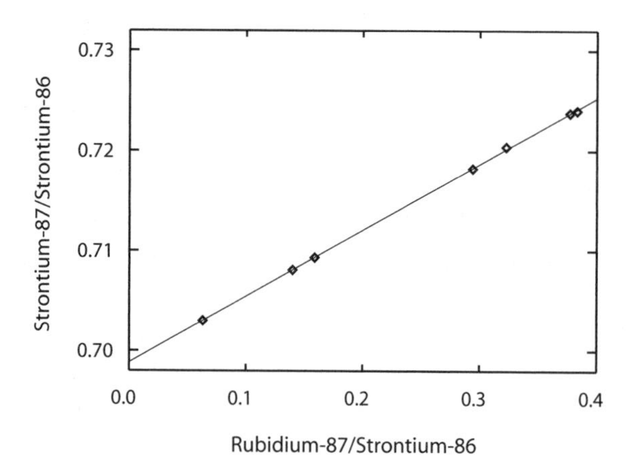

FIGURE 9.6 Isochron plot of the chondritic meteorite Tieschitz.

Finally, Figure 9.6 presents some actual data from a chondritic meteorite called Tieschitz.[2] Notice that the data fall along a line, indicating that the various components of the meteorite formed at the same time from a common source. The *y*-intercept of this line indicates that the original strontium-87/strontium-86 ratio was roughly 0.7, and the slope of the line tells us that the minerals in this body solidified roughly 4.5 billion years ago. Almost all well-dated chondrites and many achondrites have comparable age estimates, making them tens of millions of years older than even the most ancient rocks found on earth. This supports the idea that these objects are relics from the early solar system and suggests that they should be able to provide insights into the formation of solid objects in the solar system.

SECTION 9.3: SHORT-LIVED ISOTOPES AND REFINED AGES

The transformation of a disk of dust and gas around a young sun into the asteroids, comets, moons, and planets of today was a complicated process. Both computer simulations and observations of dusty disks and solar systems around other stars have contributed to our understanding of how the solar

2. Data from J.-F. Minster and C. J. Allegre "^{87}Rb-^{87}Sr Chronology of the H Chondrites..." *Earth and Planetary Science Letters* 42 (1979): 333–347.

system formed. Meteorites—especially chondrites—can also play a particularly important role in this area because they are so ancient.

Chondritic meteorites are basically agglomerations of many millimeter-sized blobs of material like chondrules that are held together in a rocky matrix. Many researchers suspect that most of the objects in the inner solar system were originally formed from such tiny bodies, as illustrated in Figure 9.7. According to this model, the chondrules and other small objects are a combination of melted

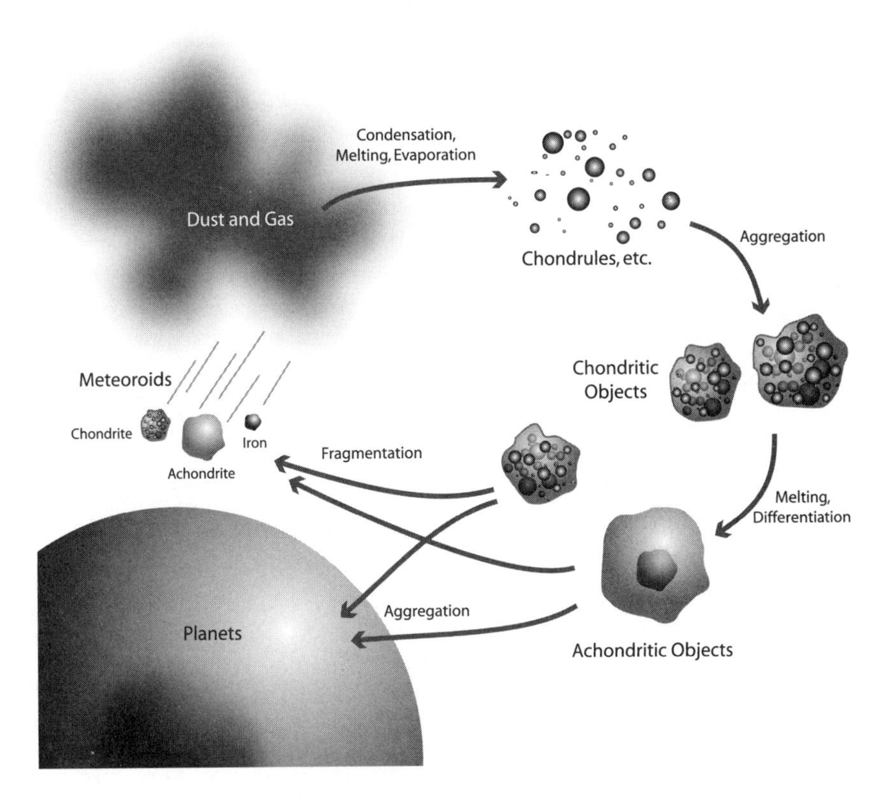

FIGURE 9.7 An illustration of the steps by which the solid material in the inner solar system could have formed. First, the dust and gas surrounding the early sun condenses and melts to form small (roughly 1 millimeter across) objects like chondrules. Next, these objects aggregate into larger and larger chondritic objects. Some of these objects melt and differentiate, destroying the chondrules within them. As these objects collide with each other, they sometimes produce fragments like meteorites, while other times they stick together. A few objects accumulate enough material to grow to the size of planets.

dust grains and condensed gases. These little bits of rock then ran into each other, occasionally sticking together to make clumps of various sizes. Some of these did not accumulate much material and remained relatively small compacted agglomerations of chondrules and dust. Others managed to accumulate more material. The heat carried into these rocks by nuclear decay and collisions could not escape efficiently from these larger bodies, which brought their internal temperature up until the rock was partially or completely melted. This destroyed the chondritic texture of the rocks, reset their radiometric clocks, and allowed materials with different chemical properties to redistribute themselves throughout the body, yielding a metal-rich core and a silicate-rich outer layer.

Over time, smaller rocks tended to aggregate into larger and larger objects. The larger the object, the stronger its gravitational field, so the biggest objects were able to grow rapidly and most of the material eventually wound up in a few planet-sized objects. Some material, however, remained in the form of smaller bodies such as asteroids. Collisions involving these objects could produce fragments that are sent hurtling through space to eventually crash-land on earth. Collisions involving asteroids small enough or cold enough to remain a collection of chondrules produced the chondrites, while objects that have partially or completely melted gave rise to some achondrites and iron meteorites.

It is important to realize that only objects on certain orbits can plausibly collide with earth, so the meteorites that we have to study do not necessarily give us a complete picture of the early solar system. For example, at distances far enough away from the sun, rocky material probably formed along with large amounts of solid ice. The early outer solar system, where the giant planets now reside, therefore probably had a very different history from the areas closer to us. The above model therefore may only apply to the formation of the inner solar system, where Mercury, Venus, Earth, Mars, and many of the asteroids are found.

Even though the basic sequence of events illustrated in Figure 9.7 is reasonable given the available data,[3] there are many open questions about the details of these different steps, such as: What caused the dust to melt? How did the chondrules get incorporated into asteroid-sized objects? How long

3. Note that some planetary scientists have suggested alternative scenarios in which chondrules were not formed directly from an accumulation of dust and gas, but instead arose when larger, partially molten objects collided with each other. They suggest that such collisions would produce sprays of liquid rock flying into space, which could then solidify into tiny, chondrule-like grains.

did it take for large objects to differentiate? In order to answer these sorts of questions, researchers have been examining the various components of chondritic meteorites in great detail. While chondrites as a whole can have a chemical makeup similar to that of the sun, different parts of these rocks contain minerals with a variety of structures and compositions. The minerals and mixes of elements found in chondrules are not the same as those found in the surrounding matrix, and even the chondrules themselves can contain a variety of mineral structures and a range of different isotopic and elemental compositions. Careful studies of the mineralogy and isotopic composition of these different components are providing fascinating information about the temperature and chemical environments in which these bodies formed. However, more precise dating methods can also provide insights into the sequence and duration of the events that produced both chondrites and achondrites.

The scenario illustrated in Figure 9.7 suggests that chondrules or chondrites should in general be somewhat older than achondrites. Furthermore, within chondrites, regions with different chemical properties could have condensed and congealed at different times. For example, there are irregularly shaped lumps of material in meteorites that contain large amounts of calcium and aluminum. These calcium-and-aluminum-rich inclusions or CAIs contain minerals and elements that melt at higher temperatures than the materials typically found in other parts of the chondrites. These minerals would also condense relatively rapidly from a liquid or gaseous state, so we expect that the CAIs would have formed even earlier than the chondrules and the rest of the chondritic materials.

In order to test these sorts of hypotheses, planetary scientists need a way to precisely measure the age differences between CAIs, chondrules, and achondrites. This sort of precision dating does not usually involve the rubidium-strontium method described above. The extremely long half-life of rubidium-87, which makes it useful for measuring the absolute age of ancient rocks, also means that it takes a very long time for the rubidium-87 content to change by a detectable amount. It is consequently difficult to establish whether CAIs are older than chondrules with this long-lived isotope. To clearly resolve differences in age between these objects, scientists employ radioactive nuclei with shorter half-lives. In fact, a great deal of useful information has been obtained from such short-lived isotopes as aluminum-26, which transforms into magnesium-26 with a half-life of only 730,000 years.

It might at first seem strange that an isotope like aluminum-26 can provide useful information about events that happened over 4.5 billion years ago. Certainly, all of the aluminum-26 in any CAI or chondrule has long since decayed away, so we cannot measure the aluminum-26 content in a meteorite today and calculate how long ago the rock formed. However, since the amount of aluminum-26 available when chondrules and CAIs formed is likely to change with time as the nuclei decay, in principle objects that form at different times will have different aluminum-26 contents when they solidify. Therefore, if we can measure the *initial* aluminum-26 content of these objects, we can perhaps discover which objects formed first and which formed later.

The initial amount of aluminum-26 in a rock can be measured because it leaves a detectable imprint on the magnesium and aluminum content of the rock. More precisely, the past presence of aluminum-26 leaves excess magnesium-26 in various minerals, just as the past presence of rubidium-87 leaves excess strontium-87. We have already seen that the data from different minerals allows us to find out how much of the strontium-87 in the rock was there originally and how much was due to the decay of rubidium-87. Similarly, we can determine how much of the magnesium-26 in the rock is primordial and how much derives from the decay of aluminum-26. To do this, we must compare the magnesium-26 content of multiple minerals with different concentrations of the stable, nonradiogenic isotopes of magnesium and aluminum, magnesium-24 and aluminum-27.

Imagine that we extracted a single chondrule or CAI from a meteorite, and measured the magnesium-24, magnesium-26 and aluminum-27 contents of several different minerals within it. From these data, we could calculate how much aluminum-27 and magnesium-26 there is per milligram of magnesium-24 in each mineral and plot this data on a graph. Figure 9.8 shows a few examples of the results of this sort of analysis, including data from a CAI belonging to the Allende meteorite along with the measurements of a chondrule from the Inman meteorite. Recall that the magnesium-24 and magnesium-26 have nearly identical chemical properties, so we expect that these objects should have all contained the same mix of magnesium isotopes when they formed, provided that they all formed in the same environment and obtained their magnesium from the same source.[4] By contrast, for both the Allende CAI and the Inman chondrule we find today that the more aluminum a mineral contains, the larger its magnesium-26/magnesium-24 ratio is. The magnesium content of these minerals therefore must have changed after

4. In practice, scientists must also account for the possibility of mass fractionation among the different minerals. As with the carbon-14 measurements discussed in chapter 5, this is done by comparing the magnesium-26 levels with the amounts of both magnesium-25 and magnesium-24.

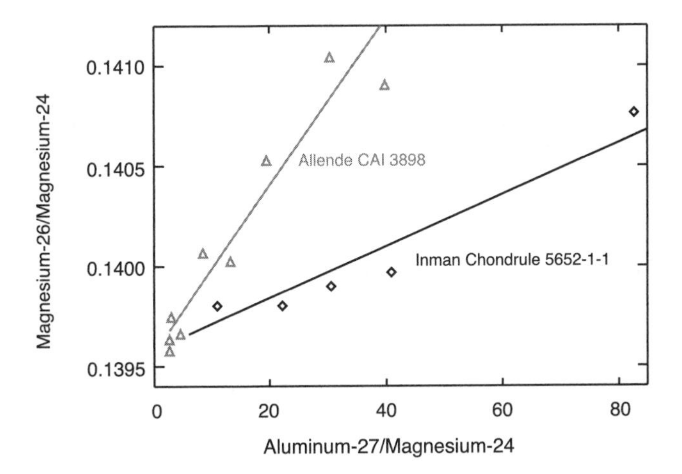

FIGURE 9.8 Isochron plots that reveal the initial aluminum-26 contents of a CAI and a chondrule. The triangles are data from a CAI belonging to the Allende meteorite and the diamonds are from a chondrule of the Inman meteorite. Data are offset slightly for clarity, and the lines drawn through the points are merely to guide the eye. Both the Allende CAI and the Inman chondrule show a correlation between magnesium-26 and aluminum-27, which suggests they both contained some aluminum-26 when they formed. The steeper slope of the CAI data implies it contained more aluminum-26 than the chondrule, and thus likely formed earlier.

these objects solidified. Since the magnesium-26 content is correlated with the aluminum-27 content, it is most likely that the decay of aluminum-26 is responsible for the altered magnesium ratios. Imagine that when the minerals in these objects first formed, they contained both aluminum-26 and aluminum-27. Again, the chemical properties of these two isotopes of aluminum are nearly the same, so minerals that have more aluminum-27 today would have also had more aluminum-26 when they solidified. After a few million years, this aluminum-26 converted into magnesium-26, increasing the magnesium-26/magnesium-24 ratio in the samples. The minerals with a higher aluminum content then would have acquired more magnesium-26 from the decay of aluminum-26 than minerals with lower concentrations of aluminum, exactly as we observe.

As with the rubidium-strontium isochron plot, we can calculate the original magnesium-26 and aluminum-26 content of the various minerals by fitting a line to the data. Again, the point where this line crosses the y-axis tells us the magnesium-26 content of a mineral that never contained any aluminum.

A mineral with no aluminum at all would not receive any magnesium-26 from radioactive decay, so such a mineral would still have the same magnesium-26/magnesium-24 ratio it had when it first formed. Assuming all of the minerals from each rock formed in the same environment, this should also be the original magnesium-26 content of all of the minerals. In this example, it turns out that the two objects originally had a magnesium-26/magnesium-24 ratio of about 0.139. Today, however, the CAI clearly contains more magnesium-26 than the chondrule. It must then have contained more aluminum-26 when it formed. In fact, for any given value of the aluminum-27/magnesium-24 ratio, the CAI has over four times as much excess magnesium-26 as the chondrule, so the CAI must have originally contained four times as much aluminum-26 as the chondrule.

If we now assume that these two objects formed in the same environment at different times, then the difference in the aluminum-26 content of the two rocks allows us to estimate the differences in the ages of these objects. To see how this works, imagine the CAI and the chondrule coalesced at different times from a cloud of dust and gas containing aluminum-26. As the millennia pass, the aluminum-26 level of the cloud steadily falls, leaving less aluminum-26 available to be absorbed into newly forming solid rocks. Since the CAI originally contained more aluminum-26 than the chondrule, the CAI must have formed earlier, when the aluminum-26 content of the cloud was higher. Furthermore, the chondrule originally contained a quarter as much aluminum-26 as the CAI, so the aluminum-26 content of the cloud must have dropped by a factor of four between the times when the CAI and the chondrule formed. Since the half-life of aluminum-26 is about 730,000 years, the chondrule must have solidified a little over 1.5 million years after the CAI did.

Measurements of the initial aluminum-26 content of CAIs and chondrules reveal very interesting patterns. Typical CAIs, like Allende's CAI, originally had about forty-five micrograms of aluminum-26 for every gram of aluminum. By contrast, most of the chondrules cosmochemists have studied originally had less than twenty micrograms of aluminum-26 per gram of aluminum. Finally, the original aluminum-26 content of several achondrites has been found to be extremely low (only a few parts per million). These data indicate that the CAIs were among the first objects to solidify in the early solar system, which is certainly reasonable given their refractory chemical makeup. Recently, some chondrules with aluminum-26 levels comparable to CAIs have been discovered, implying that at least a few chondrules started to form about the same time as CAIs. However, the lower aluminum-26 contents of many chondrules suggests that they continued to appear for a few million years afterwards. Finally, the minute, almost undetectable aluminum-26 levels of some achondrites implies that they formed later still, possibly by the reprocessing of chondrules. This sequence of events is roughly consistent with the model of solid body formation illustrated in Figure 9.7.

Of course, these age estimates assume that all of these objects obtained their aluminum-26 from a common source distributed uniformly throughout the early solar system. Furthermore, they assume that the aluminum-26 content of that source decayed steadily with time and was never replenished. This begs the question of where this aluminum-26 came from and why was it present just when CAIs were forming. One possible source of the aluminum-26 is the sun itself. High-energy particles created by a young star could produce some aluminum-26 when they collide with the gas and dust in the surrounding cloud. If this was the case, then there would be a steady flow of aluminum-26 into the solar nebula, and we should not consider aluminum-26 to be a reliable method of measuring time. In fact, if our sun was the source of the aluminum-26, then the differences between CAIs and chondrules would tell us more about *where* these objects formed than *when* they formed. For example, it could be true that CAIs formed closer to sun, where the aluminum-26 levels were high, and chondrules formed further out. However, evidence from other short-lived isotopes (like manganese-53) suggest that this explanation is somewhat unlikely. Each of these isotopes can be used to estimate when these objects formed, and they all paint a similar picture of the sequence of events that occurred in the early solar system. Furthermore, the distribution of these isotopes in chondrules and CAIs seems to be inconsistent with the predicted energetic particle radiation from the sun.

Another explanation for the aluminum-26 and other short-lived nuclei in the early solar system is that they came from outside the solar system in single burst, perhaps due to the explosion of a relatively nearby star—a phenomenon known as a supernova. This theory is attractive for two reasons. First, the particles thrown through space by such an explosion could very plausibly seed the early solar system with aluminum-26 and other unstable nuclei. Second, the shock waves associated with the supernova could have disturbed a cloud of dust and gas, causing it to collapse and triggering the formation of the solar system at about the same time as it was being doused with aluminum-26. In this scenario, the age estimates based on these short-lived radioactive elements are likely to be basically correct. However, even if a supernova did produce most of the aluminum-26 in the early solar system, it is still possible the sun also added some aluminum-26 into the mix. There is also still a chance that the aluminum-26 was not evenly distributed in the solar system. These issues could complicate our efforts to understand the early solar system.

Happily, thanks to advances in geochemistry, it is now possible to test the assumption that aluminum-26 as other nuclei were distributed evenly throughout the solar system. Recently, a dating method similar to the rubidium-strontium approach has been refined to the point that it can produce a reliable and extremely precise measurement of how long ago a chondrule or CAI formed. This method uses that most famous of radioactive isotopes, uranium.

SECTION 9.4: FINE SCALE ABSOLUTE DATES WITH THE URANIUM-LEAD SYSTEM

As far as elements go, uranium has an extremely heavy nucleus, with 92 protons and well over 100 neutrons. There are two common isotopes of uranium found in nature, uranium-235 and uranium-238. Both are unstable and decay through a complex series of steps into two different isotopes of lead. uranium-235 decays into lead-207 with a half-life of 700 million years; uranium-238 decays into lead-206 with a half-life of 4.5 billion years.

The half-lives of the uranium isotopes are long enough that some amount of uranium should persist in chondritic meteorites to the present day. Because of this, we could do an analysis of the uranium and lead content of various minerals in a material (analogous to the previous analysis of the rubidium-strontium content) to obtain an age estimate for the material. Since the half-lives of the uranium isotopes are considerably shorter than that of rubidium-87, the uranium-lead data should produce a more precise estimate than the rubidium-strontium data. However, we can do even better than this. Since there are two different isotopes of uranium that decay into two different isotopes of lead with two different half-lives, the lead content of the rock alone can provide us with an estimate of when the material formed.

Let us again imagine a newly formed meteorite composed of several different minerals. We extract minerals from this rock and measure the amount of lead-206, lead-207, and lead-204 in each one. Remember that isotopes of the same element have nearly identical chemistry, so they are picked up in the same ratios by any mineral within the rock. The various samples will therefore have the same mix of isotopes, and if we made a plot of the amount of lead-206 and lead-207 per gram of lead-204 in each sample, all of the data points would fall on the same place, as shown in the upper left panel of Figure 9.9.

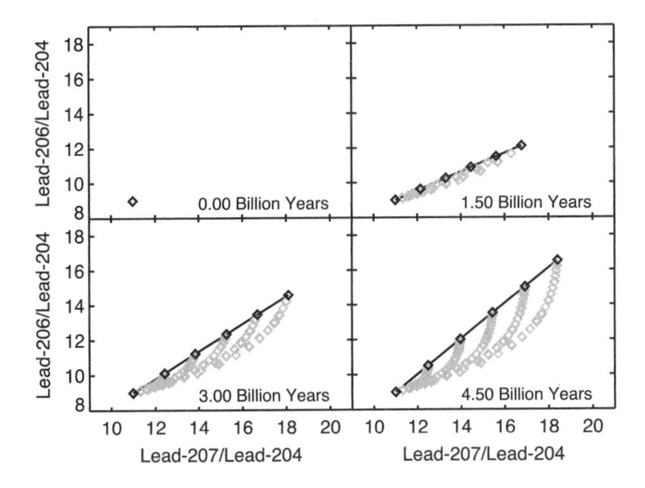

FIGURE 9.9 Lead-lead isochron dating. Each panel shows the mix of lead isotopes in a hypothetical rock at various times after its formation. The light gray points trace how the isotopic composition of each mineral has changed over time. Note that the slope of the line steadily increases with time.

This simple situation will not last, however, because these minerals also acquired some amount of uranium, and over time the uranium will decay into lead and alter the isotope ratios of the different minerals. lead-204 is not produced by nuclear decay, so the amount of it in each sample will remain constant. Meanwhile, the lead-206 and lead-207 levels will steadily increase as uranium-238 and uranium-235 decays. Since uranium-235 and uranium-238 are also isotopes of the same element, all of the minerals originally contained the same mix of these two nuclei. As the uranium decays, all minerals will acquire the same proportions of new lead-206 and lead-207. However, the total amount of uranium will be different in different samples, so the total amount of lead-206 and lead-207 produced varies from sample to sample.

Say we take another look at the minerals 1.5 billion years after they formed. This would give us a plot like the one shown in the upper right panel of Figure 9.9. The data points fall along a straight line because all of the minerals produce the same number of grams of lead-207 per gram of lead-206. The slope of this line is shallow because uranium-235 has a shorter half-life than uranium-238, so the amount of lead-207 in any sample increases faster than the amount of lead-206. However, as time goes on, the line becomes steeper because more and more uranium-238 decays, as shown in the lower panels of

Figure 9.9. The slope of this line therefore provides a way to estimate when the rock formed, just as it did for the rubidium-strontium system described above. However, an age estimate based on the lead content of the rock can in principle be much more reliable and precise than one based on a rubidium-strontium analysis. Not only are the relevant half-lives shorter, the measurements themselves are less likely to be affected by systematic errors. An age estimate based on the rubidium-strontium system requires us to determine the amounts of two different elements in each sample. Since these elements have distinct chemical properties, one has to ensure that the measuring equipment does not detect rubidium more efficiently than strontium or vice versa. By contrast, with the uranium-lead system, one need only measure the relative amounts of three isotopes of a single element. This is in principle a simpler task, so age estimates provided by this process can be extremely precise.

In 1992 an achondrite was measured to have an age of 4.558 billion years, with an uncertainty of only 500,000 years. Then, in 2002, another team measured the age of chondrules from one chondrite as 4.564 billion years and the age of two CAIs from another chondrite as 4.567 billion years, all with uncertainties significantly smaller than one million years. These data—which are impressively exact for objects so ancient—confirm that some chondrules formed a few million years after the CAIs, and that achondritic materials formed even later still. This means that the chronology based on short-lived nuclei is basically correct, and a large amount of aluminum-26 was probably deposited in the early solar system in a single burst. This lends support to the notion that a nearby supernova may have played a pivotal role in the early history of our solar system.

The data from short-lived isotopes and the uranium-lead system—together with a variety of other dating methods and careful studies of the mineral structures present in objects like CAIs and chondrules—helps us understand what our solar system was like 4.5 billion years ago. For example, the fact that CAIs and chondrules did not all form at the same time means that they did not appear following a one-time event like the aforementioned supernova. Instead, the processes responsible for creating these objects must have operated for millions of years: shock waves in the disk, solar flares, collisions, and even nebular lightning are a few possibilities being considered.

There is still much work that needs to be done before we will have a solid understanding of how a cloud of dust and gas became a collection of solid objects. For example, at present we have age estimates for only a limited number of chondrules and CAIs. This is not only because these measurements are time-consuming, but also because the rocks need to have detectable amounts of aluminum or uranium. As these techniques are refined, cosmochemists have been able to date more and more pieces of chondritic meteorites. This will not only serve to confirm or deny the above chronology, but also help us understand how these millimeter-sized particles were incorporated into larger bodies. According to some researchers, the gas in the early solar system should have slowed down small objects like chondrules and CAIs and sent them spiraling into the sun. It is not clear how CAIs could have avoided the sun for millions of years while the chondrules formed. Perhaps CAIs found refuge in larger objects as the chondrules developed, or perhaps winds and outflows from the early sun kept these bodies aloft for a few million years until they were all swept up into primitive asteroids and proto-planets. By measuring the age distributions of CAIs and chondrules from individual meteorites, we will hopefully one day know which of these scenarios is closer to what actually happened.

Progress is already being made in this endeavor. In 2004, a report described the aluminum-26 contents of CAIs and chondrules extracted from a single meteorite. In this rock, the CAIs all had almost the same initial aluminum-26 contents, roughly 50 parts of aluminum-26 per million parts of aluminum. The chondrules, by contrast, have a range of initial aluminum-26 contents: from 50 parts per million to less than 20 parts per million. This implies that this meteorite contains grains with a range of ages. Even more recently, in 2005, another team announced that certain achondrites may be as old as chondrules, implying that some objects were differentiating very early on. As more dates are extracted from meteorites in the coming years, researchers should be able to assemble these pieces of information into a much clearer picture of the origins and early history of our solar system.

Of course, there are many other star systems besides our own scattered far and wide throughout the galaxy, and each one has a story to tell. Since we can only observe them from a distance, obtaining detailed information about stars beyond the sun poses special challenges, but careful inspection of the light from certain stars has provided valuable clues about both their histories and their ages.

SECTION 9.5: FURTHER READING

For details about the Park Forest meteorite, try S. B. Simon et al. "The Fall, Recovery, and Classification of the Park Forest Meteorite" *Meteoritics and Planetary Science* 39, no. 4 (2004): 625–634. For a nice introduction to meteorites, see Harry McSween *Meteorites and Their Parent Planets,* 2nd ed. (Cambridge University Press, 2000) and J. Kelly Beatty, Carolyn C. Petersen, and Andrew Chaikin *The New Solar System,* 4th ed. (Cambridge University Press, 1999) and Paul Weissman, Lucy-Ann L. McFadden, and Torrence Johnson *Encyclopedia of the Solar System* (Academic Press, 1999).

For more technical overviews of meteorites, try Vincent Mannings, Alan P. Boss, and Sara S. Russell eds. *Protostars and Planets IV* (University of Arizona Press, 2000); A. M. Davis ed. *Meteorites, Comets, and Planets*, vol. 1 of *Treatise on Geochemistry*, H. D. Holland and K. K. Turekian, general editors (Pergamon, 2004); and Robert Hutchinson *Meteorites* (Cambridge University Press, 2004).

For more details on isochron plots and radiometric dating, look at textbooks on geochemistry such as A. H. Brownlow *Geochemistry,* 2nd ed. (Prentice Hall, 1996). There is also a good discussion on the web at www.talkorigins.org/faqs/isochron-dating.html.

For the data used to make Figure 9.8, see Frank A. Podosek et al. "Correlated Study of the Initial $^{87}Sr/^{86}Sr$ and Al-Mg Isotopic Systematics and Petrologic Properties in a Suite of Refractory Inclusions from the Allende Meteorite" *Geochimica et Cosmochimica Acta* 55 (1991): 1083–1110 and S. S. Russell et al. "Evidence for Widespread 26Al in the Solar Nebula" *Nature* 273 (1996): 757–762. A good review of aluminum-26 dating can be found in G. J. MacPherson et al. "The distribution of aluminum-26 in the early solar system—a reappraisal" *Meteoritics* 30 (1995): 365–386.

For the latest high-precision age estimates using the uranium-lead system, see G. W. Lugmair and S. J. G. Galer "Age and Isotopic Relationships among the Angrites" *Geochemica et Cosmochemica Acta* 56 (1992): 1673–1694; Yuri Amelin et al. "Lead Isotopic Ages of Chondrules and Calcium-Aluminum-Rich Inclusion" *Science* 297 (2002): 1678–1683; Yuri Amelin et al. "Unraveling the Evolution of Chondrite Parent Asteroids by Precise U-Pb Dating and Thermal Modeling" *Geochimica et Cosmochimica Acta* 69 (2005): 505–518; and Joel Baker et al. "Early Planetesimal Melting from an Age of 4.5662 Gyr for Differentiated Meteorites" *Nature* 436 (2005): 1127–1131.

Some recent studies of the aluminum-26 contents of Chondrules and CAIs are found in Martin Bizzarro et al. "Mg Isotope Evidence for Contemporaneous Formation of Chondrules and Refractory Inclusions" *Nature* 431 (2004): 275–278, and Alexander N. Krot et al. "Chronology of the Early Solar System from Chondrule-Bearing Calcium-Aluminum-Rich Inclusions" *Nature* 434 (2005): 998–1001.

For those interested in the current state of research on meteorites and early solar system formation, a good place to start is the website of the Lunar and Planetary Institute, http://www.lpi.usra.edu/resources.

CHAPTER TEN

Colors, Brightness, and the Age of Stars

On February 24, 1987, the world was treated to a rare astronomical show. In less than a single day, a star that had previously been observable only with a telescope suddenly increased in brightness by a factor of a thousand, becoming visible to the naked eye. The object then slowly faded away over the following months. This event, called Supernova 1987A, dramatically demonstrates that stars are not eternal celestial jewels. In fact, no star can be expected to shine forever.

Every star, including our sun, is an enormous ball of (mostly) hydrogen gas held together by gravity and illuminated by the nuclear reactions occurring deep in its core. Eventually, the nuclear furnace of any star will run out of fuel and it will lose its ability to generate large amounts of light. The stars we see around us today therefore cannot be infinitely old. Instead, every star must have formed at a definite, measurable time in the past. Many astrophysicists even think that there was a time long ago when there were no stars shining at all. By measuring the ages of different stars, astronomers can help establish when the first stars started to light up the universe.

SECTION 10.1: STUDYING STARLIGHT

One of the biggest challenges astronomers face in studying many stars are the enormous stretches of space separating them from our solar system. Even the closest stars beyond our solar system are over thirty trillion kilometers away. To get a sense of just how large this distance is, consider the following: light travels at the remarkable speed of almost 300,000 kilometers per second, and can cover the entire distance from the sun to the earth in under ten minutes, yet it takes several years for light from the sun to reach the closest stars! The stars of the night sky are so far away that they almost always appear as simple points of light in even the most powerful telescopes. This means that we cannot usually observe the shape or size of a star directly, much less discern features on its surface.

In spite of these limitations, astronomers are able to extract a considerable amount of data about stars' characteristics from the light they produce. Much of this information is encoded in the stars' spectra, which are revealed by breaking the starlight into its component colors. This could be done using a prism, but in practice devices like diffraction gratings and interferometers are better choices. In either case, the light passing through the device forms the familiar rainbow of colors extending from red to violet and beyond. This happens because light is an electromagnetic wave, and the wavelength of the light affects how it interacts with material objects like a prism or diffraction grating. Light with different wavelengths leaves the device in different directions, so red light—which has a longer wavelength—separates from the blue light, which has a shorter wavelength. Once the light is dispersed in this way, we can quantify the spectrum of the star by measuring the brightness of the light at various wavelengths. Since the wavelength of the light is related to its color, these plots provide a generalized description of the color of the star.

Stellar spectra usually show a broad peak extending over a wide range of wavelengths (combined with a series of dips; see Figure 10.1). The width and position of this peak is different for different stars, but the basic shape is always characteristic of thermal radiation: light produced by the random motions of atoms, electrons, and nuclei. The spectrum of thermal radiation is not very sensitive to the composition or structure of the object, and instead depends primarily on its temperature. As the temperature increases, the particles move faster and faster, which results in the peak of the spectrum moving to shorter and shorter wavelengths. An object with a bluish glow is therefore hotter than one with a reddish glow. This means that the shape of the spectrum often provides a good measurement of the temperature of the visible parts of the star.

Obtaining detailed spectra is time-consuming, and it cannot easily be done for large numbers of stars in a reasonable amount of time. Astronomers doing surveys of many stars therefore sometimes prefer to estimate the spectra by measuring the total amount of light transmitted through a handful of different filters, each of which lets through only a part of the spectrum. These filters are

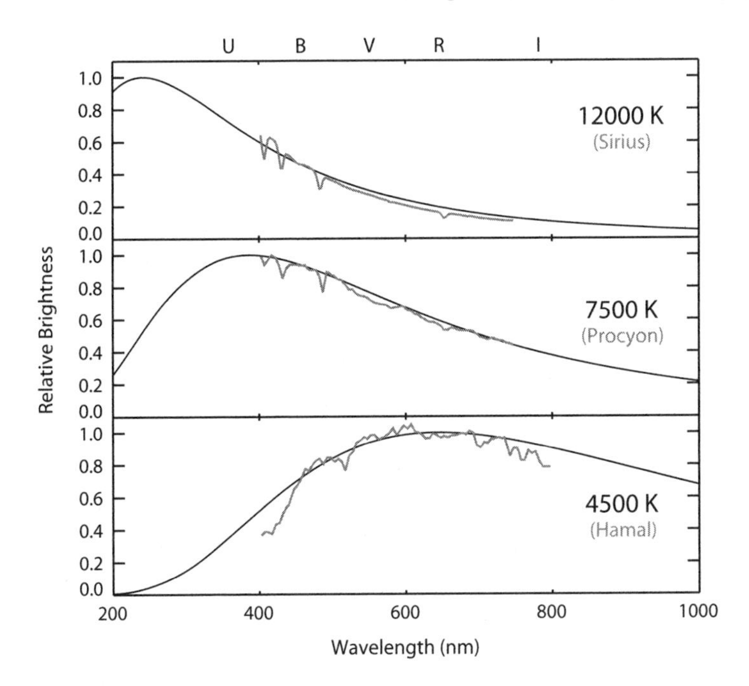

FIGURE 10.1 Examples of stellar spectra, which illustrate how the brightness of stars can vary as a function of the wavelength of the light (blue is towards the left, red is towards the right). The actual spectra of three stars (shown here in gray) have a number of dips and wiggles. However, the basic shape of the spectra is consistent with thermal emission. Thermal spectra, illustrated by the smooth curves, have a broad peak at a wavelength that depends on the temperature of the source. The labels on each curve indicate the effective temperature (in Kelvins) for each star spectrum. Note that higher-temperature objects have a spectrum that peaks at shorter wavelengths. The letters along the top of the plot represent some of the filters commonly used to measure starlight (U = "Ultraviolet," B = "Blue," V = "Visible," R = "Red," I = "Infrared"). The visible star spectra come from the Burnashev (1985) catalogue available at http://vizier.cfa.harvard.edu/viz-bin/VizieR?-source=III/126.

labeled by letters that identify the wavelengths that pass through the filter (see Figure 10.1). For example, the *B* filter transmits *blue* light, while the *V* filter transmits longer-wavelength *visible* light.

Conventionally, the amount of light transmitted through a filter is described in terms of an apparent magnitude. For most nonastronomers, magnitudes are tricky things to understand, because they are not a "unit" in the conventional sense, like meters or seconds. While a 500-meter race is five times as long as the 100-meter dash, and a three-hour concert is three times as long as a one-hour concert, a 4th magnitude star is *not* twice as bright as a 2nd magnitude star. Instead, a star with magnitude 1 is about 2.5 times brighter than a star with magnitude 2, which is in turn 2.5 times as bright as a star with magnitude 3. Notice that stars with larger magnitudes are fainter. Also, the *difference* in the magnitudes of two stars indicates the *ratio* of their brightness. This means that a magnitude 23 star is 2.5 times brighter than a magnitude 24 star, just like a magnitude 1 star is 2.5 times brighter than a magnitude 2 star. While this system of measuring brightness might at first seem odd and confusing, it does have some nice features for astronomers. In particular, a factor of a million in brightness corresponds to a difference of only 15 magnitudes, so large ranges of brightness can be considered without having to keep track of very large or very small numbers.

The magnitude of a star measured through a particular filter depends on its spectrum. For example, a relatively cold star (the lowest curve in Figure 10.1) produces more light in the *V* part of its spectrum than in the *B* part, so the *B*-band magnitude is higher than the *V*-band magnitude for this star. By contrast, hotter stars produce more light in the *B* part of the spectrum, so their *B*-band magnitudes can be lower than their *V*-band magnitudes. The difference in the magnitudes *B-V* then provides a crude measurement of the shape of the spectrum and the temperature of the star. In astronomical parlance, these differences are known as colors. Keep in mind that because a lower magnitude corresponds to more light, a *smaller* value of *B-V* means that the star is *bluer* and therefore *hotter*.

If interstellar space were a perfect vacuum the color of a star would not depend on how far away it is: a star 100 light-years away would look just as red if it were 200 light-years away.[1] By contrast, the overall brightness of a star is directly related to its distance from us. As the light travels away from the star, it spreads out over a larger and larger area, making it progressively fainter. However, just because one star in the sky appears less bright than another, it does not necessarily follow that this star is more distant. Stars can also appear brighter or fainter simply because they produce more or less light. For example, the bright star the Egyptians used to mark the beginning of the New Year—today known as Sirius—has a companion star in orbit around it. Both of these stars are the same distance away from us, but Sirius is over

1. A light-year is a common astronomical unit of distance. It is the distance light can travel in a year, and it corresponds to approximately 9.5 trillion kilometers. Note that because of obscuration by interstellar dust, the color of a star sometimes does depend on its distance away from us.

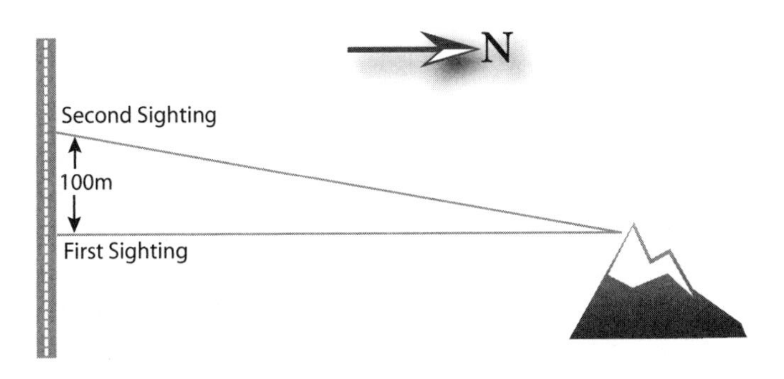

FIGURE 10.2 Using parallax to measure distances on earth. The angle between the two sight-lines and the distance between the sightings provides sufficient information for us to calculate the distance to the mountain (illustration not to scale).

10 thousand times as bright as its companion. These two stars are therefore generating very different amounts of light.

Like the shape of its spectrum, the total amount of light produced by a star also provides important clues about its physical properties. However, we can only calculate this parameter after we have figured out how far away the star is. Because stars are so remote, astronomers have had to develop a variety of clever techniques to measure their distances. The most direct method begins by measuring the position of the star in the sky at a few different times over a year. During each of these observations, the earth will be at a different point in its orbit around the sun, so the star will appear at slightly different positions. This phenomenon, known as parallax, allows us to use simple trigonometric formulas to discover the distance to the star.

To see how this method works, consider a similar technique used to measure distances here on earth. Imagine we were on an east-west road. At one point on the road we find a distant mountain peak was due north of us. Then, suppose we walked 100 meters west and found that from our new location the mountain peak is one degree east of north. The mountain peak is located at the intersection of the two sight-lines, which means that the mountain and our two observing points form a triangle (Figure 10.2). The distance between the two observations tells us the length of one side of the triangle (100 m), and the angles between the sides can be extracted from the apparent position of the mountain. This provides us with enough information to determine the lengths of all three sides of the triangle and to work out that the mountain lies about 6 kilometers away. By the same logic, astronomers know how far the earth moves between observations weeks or months apart, so they can use similar calculations to determine the distance to a star (see Figure 10.3).

While parallax provides a very direct measurement of stellar distances, it does have limitations. Think of how, when you watch the scenery from a train, the nearby trees pass by quickly while the hills in the distance appear to move much more slowly. By the same token, the apparent motion of a star is less if it is farther away. If the stars are beyond a certain distance away, they do not appear to move by any detectable amount and this method cannot be used. Parallax can therefore only provide us with the distances to relatively nearby stars. However, over the decades astronomical instruments have improved, enabling parallactic measurements of more distant stars. In the 1990s, the Hipparcos satellite measured the distances to tens of thousands of stars, some of which are about a thousand light-years away.

Once we have the distance to a star, we can work out how much total light it has to produce to account for its observed brightness here on earth. Often, this quantity—known as the star's luminosity—is presented relative to the luminosity of our sun. Sirius, for example, is about 23 times solar luminosity. Its companion, by contrast, is only 0.002 times solar luminosity. Strictly speaking, the luminosity of a star is the total amount of light it produces at all wavelengths, but it is often impractical to measure this directly, so the luminosity is instead estimated based on the total light transmitted through various filters. These signals are usually described in terms of a magnitude scale. However, in this case the *apparent* magnitudes of stars we observe on earth are converted to *absolute* magnitudes using the relevant distance information. Astronomers define the absolute magnitude of a star as the magnitude it would have if it were located 32.6 light-years away. For example, the apparent V-band magnitude of Sirius is about −1.5, and it is located about 9 light-years away. If it were 32.6 light-years away, it would be about 12 times fainter, and its absolute magnitude works out to +1.5. For comparison, the absolute magnitude of the sun is about +4.8.

Studying the light from a star can provide us with more than just its luminosity and temperature. High-resolution star spectra show a series of dips indicating the presence of specific atoms and molecules in the stellar atmosphere. Careful studies of these spectral lines can tell us something about the composition of the star, and changes in the locations of these features are used to reveal the presence of nearby planets. However, the absolute magnitudes and colors will be sufficient for our purposes here. These parameters are tightly coupled to the inner workings of the stars, so they provide important clues about stars' internal characteristics and history.

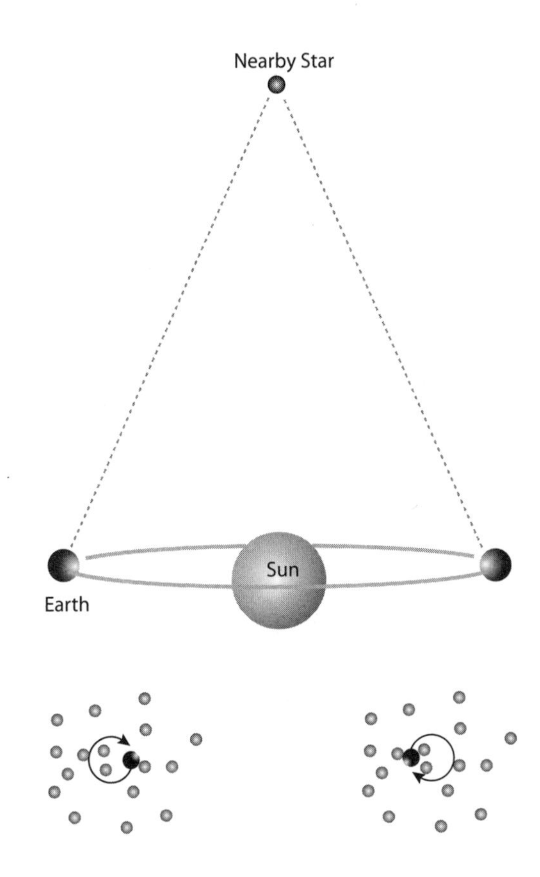

FIGURE 10.3 Using parallax to measure the distance to stars. As the earth moves around the sun over the course of a year, an observer on earth will view a nearby star from slightly different angles and the star will appear at slightly different positions in the sky. In practice, the changing position of the star is determined based on its location relative to more distant stars, which do not move as much. The apparent motion of the nearby star over the year is illustrated by the panels at the bottom of the figure. Astronomers can then use this movement along with some basic trigonometry to calculate the distance between the earth and the star.

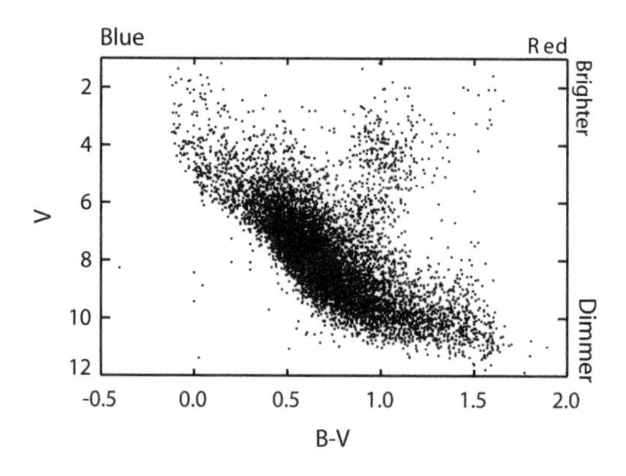

FIGURE 10.4 A color-magnitude diagram of nearby stars, based on data from the Hipparcos satellite, available at http://vizier.cfa.harvard.edu/viz-bin/VizieR?-source=I/239. This is a plot of absolute magnitude in *V*-band versus the color *B-V*, with each point representing a single star. Points that lie towards the left of the plot are bluer than stars that lie to the right, and stars towards the top are more luminous than stars nearer the bottom. Most of the stars fall along a diagonal line known as the main sequence.

Different types of stars have different patterns in their luminosity and their surface temperatures, which are often illustrated using a color-magnitude diagram (also known as a Hertzsprung-Russel diagram). Such diagrams show the (absolute) magnitude of a collection of stars versus their color. The location of a point along the horizontal axis indicates a single star's color and temperature: hotter, bluer stars are to the left and cooler, redder stars are to the right. The point's location along the vertical axis corresponds to the star's absolute magnitude, with more luminous stars higher up and less luminous stars lower down. Figure 10.4 is a color-magnitude diagram of some 10,000 nearby stars with distances measured by the Hipparcos satellite. Note that almost all of the points fall along a fuzzy diagonal band extending from the upper left to the lower right of the plot. This feature is known as the main sequence, and it corresponds to a particular class of stars with a specific relationship between surface temperature and luminosity: the hotter (bluer) stars tend to be more luminous than colder (redder) stars. This trend occurs because all of these stars share certain fundamental characteristics. Specifically, they all appear to generate energy primarily through the fusion of hydrogen into helium.

SECTION 10.2: THE LIVES OF MAIN SEQUENCE STARS

Hydrogen is a natural energy source for most stars because hydrogen is the most abundant element in the universe and forms the bulk of most stars. Even though normal hydrogen is the simplest element—with a single proton in its nucleus—these atoms are still able to generate sufficient light and heat to fuel a star because they can be assembled or fused together into helium nuclei.

Figure 10.5 shows one of the methods by which four protons become one helium nucleus. First, two protons form a nucleus of deuterium, a heavy form of hydrogen with one proton and one neutron. Since one proton converts into a neutron during this process, a positron, or anti-electron, and a neutrino are emitted. Next, another proton combines with this deuterium nucleus to

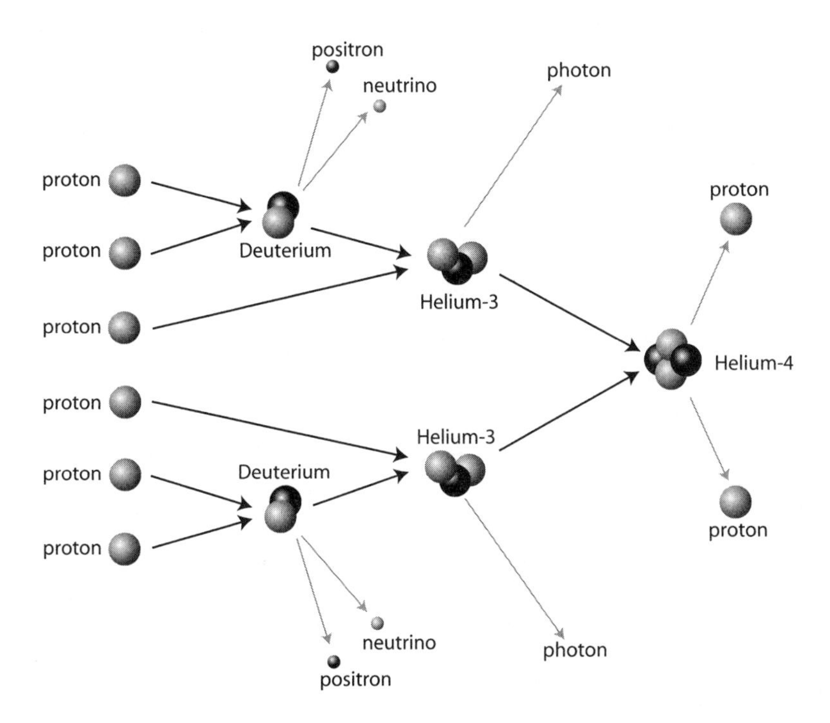

FIGURE 10.5 The fusion of hydrogen into helium. Hydrogen nuclei (the individual protons on the left) can be assembled into a helium nucleus in a variety of ways. The process illustrated here is the most straightforward (other process use heavier nuclei as catalysts). Each of the reactions illustrated here reduces the mass of the nuclei, releasing energy and provoking the motion of the various particles.

form a variant of the helium nucleus with two protons and one neutron. To conserve energy and momentum, a photon is emitted during this step. Finally, two of these light helium nuclei come together, throw off two protons, and leave behind a normal helium nucleus, with two protons and two neutrons.

This process is capable of powering a star because the mass of a helium nucleus is about 0.7% less than the total mass of four hydrogen nuclei. Remember, Einstein's famous equation $E = mc^2$ says that there is an energy associated with massive objects. This means that when hydrogen nuclei are fused together, some of the mass-energy in the nuclei converts into other forms of energy, such as electromagnetic radiation and particle motion. While 0.7% may not sound like an impressive figure, it actually corresponds to a huge amount of energy. A single gram of hydrogen converted into helium releases as much energy as burning 20 metric tons of coal. Hydrogen is obviously an extremely potent power source, but energy can be extracted efficiently from these nuclei only under the right conditions. The hydrogen nuclei must get extremely close to each other before any nuclear reactions can occur, and since hydrogen nuclei are all positively charged, they repel each other. Fusion will therefore occur only if the nuclei are pressed close together or if they collide at very high speeds. These sorts of conditions have not yet been achieved on earth without using more energy than the reactions produce, which is why fusion power plants are not available yet. However, stars contain so much hydrogen and have such immense gravity that these nuclear reactions occur naturally.

Imagine a diffuse cloud of hydrogen and helium gas in space that contains about as many atoms as our sun. Gravity pulls every atom towards every other atom, drawing them all down into the center of the cloud. As the cloud begins to collapse in on itself, more and more gas accumulates in the core, causing the mass in this region to rise. The flow of material into the middle of the cloud therefore accelerates until enough hydrogen atoms are crammed into a small enough space with a high enough temperature for nuclear fusion to occur. The cloud of gas then begins to transform into a young star. The nuclear reactions in the center of this object produce radiation and fast-moving particles that fly out of the core to collide with the material being dragged inwards by the force of gravity. As the density of the core continues to increase, the nuclear reactions occur more and more rapidly. Eventually, the outward push generated by the fusion in the core balances the inward pull of gravity and the star reaches a state of equilibrium.

Once a star settles down and reaches equilibrium, it can remain nearly in balance for as long as it is able to fuse hydrogen in its core. If this fusion becomes insufficient to support the star, gravity will bring more material into the core, causing the nuclear reaction rate to increase until the star stops collapsing. Conversely, if the energy released by fusion reactions was ever more than enough to support the star, material would be blown outward, reducing the pressure on the core, and the nuclear reaction rate would drop until the star stabilized. We therefore might expect that most stars—those that fall along the main sequence—exist in just such an equilibrium state.

If main sequence stars really are close to equilibrium, then the region outside their cores should be nearly in a steady state. Different atoms in the star may move inwards or outwards at different times, but on average there is almost no net flow of material towards or away from the core, and the average speed of particles in any part of the star remains constant. As a result, the outer layers of these stars neither gain nor lose energy, even though fusion reactions in the core are releasing vast quantities of energy. Given that energy cannot be created or destroyed, the light and other forms of radiation emitted by a star have to carry energy away from the surface as fast as it is produced in the core. The characteristics of starlight—especially the luminosity—should therefore be tightly coupled to the rate of nuclear reactions in the core. Furthermore, since the core's fusion rate must be sufficient to balance the force of gravity, both the surface temperature and the luminosity of main sequence stars should be strongly correlated with their total mass.

Astronomers have confirmed that such a connection between mass, luminosity, and surface temperature does indeed exist in main sequence stars. About half of all nearby stars are in binary systems, where two stars orbit around each other. Astronomers can actually watch the stars move around in their orbits over the years. Just as the time it takes the earth to go around the sun depends on the sun's mass, the time it takes for these stars to complete an orbit depends on their masses. Using this information, astronomers have been able to determine the masses of hundreds of main sequence stars. These data, some of which are shown in Figure 10.6, clearly show that there is indeed a direct relationship between the mass of a main sequence star and its luminosity: the more massive a star is, the more luminous it is. If we chose to make a graph of mass versus surface temperature, we would find a similar result: more massive stars are also bluer and hotter than less massive stars. These relationships are what we would expect from stars in an equilibrium state. The more massive a star is, the faster the fusion rate has to be to prevent the star from collapsing under its own gravity. This higher rate of energy production yields a higher luminosity while simultaneously raising the temperature of the star's surface.

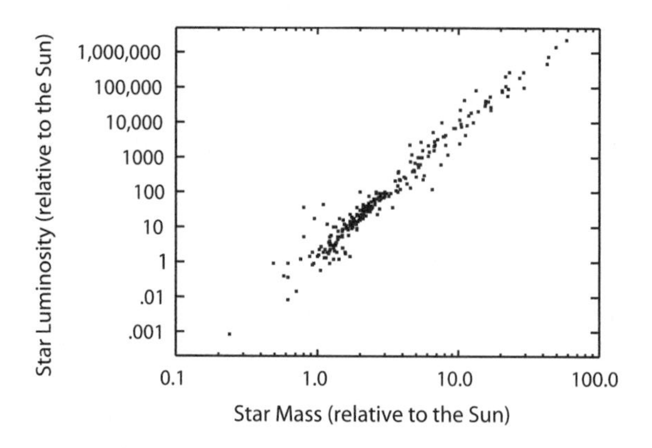

FIGURE 10.6 The mass-luminosity relation, illustrated by a plot of the stellar luminosity versus mass. Note that when the mass increases by a factor of 10, the luminosity increases by a factor of several thousand. (Based on data from Svechnikov and Bessonova (1984) available at http://vizier.cfa.harvard.edu/viz-bin/VizieR?-source=V/42).

The observed mass-luminosity relationship not only documents a relationship between fundamental parameters in main sequence stars, it also indicates that the mass of a main sequence star has a disproportionately large affect on its luminosity. A star 10 times as massive as our sun is not merely 10 times as luminous as the sun: it emits 10,000 times as much light! This occurs because more massive stars not only have more material to support, they also have stronger gravitational forces pulling on every single particle. In addition, the particles in a heavier star must move more rapidly to maintain themselves aloft, and this in turn increases the rate at which energy is carried through the star and lost into space. Increasing the mass of a star therefore causes its power requirements to rise very quickly, and this has important implications for stars' life spans and ages.

SECTION 10.3: THE DEATHS OF MAIN SEQUENCE STARS

By definition, a star that is in equilibrium does not change appreciably over time. This means that a star with a given color or luminosity may have only recently reached equilibrium or it may have been burning hydrogen for billions of years. However, main sequence stars cannot last forever because they all eventually run out of nuclear fuel. Astronomers can therefore extract some chronological information from stars if they can determine how long they can live on the main sequence.

The life span of a main sequence star is set by how long the fusion reactions in its core are able to release enough energy to support it. Remember that nuclear reactions release energy only if the total mass of the particles after the reaction is *less* than that of the particles going into the reaction. Most reactions that involve helium-4 nuclei *increase* the mass of the nuclei involved and cannot help keep the star in equilibrium.[2]

A star can use helium-4 nuclei as a power source only if it can assemble them into carbon-12 nuclei. A carbon-12 nucleus has as many protons and neutrons as three helium nuclei, and is less massive overall than these nuclei, so this reaction releases energy that can used to support the star. However, this process requires three helium nuclei to be in nearly the same place at nearly the same time, so helium fusion will only occur at much higher temperatures and densities than hydrogen fusion. In a main sequence star, the energy released from hydrogen fusion keeps the material in the core from collapsing into such a dense, hot state. This means that the fusion of hydrogen to helium is the only practical energy source for a main sequence star, and so these stars will be unable to maintain their equilibrium after they use up too much of their hydrogen.

Even though about 90% of the atoms in a young main sequence star are hydrogen, only a fraction of this material can be used as fuel because hydrogen fusion occurs only in the densest central parts of the star. Since helium nuclei are more massive than hydrogen nuclei, the helium generated within the core is not efficiently carried into the outer layers of the star. Instead, it accumulates in the depths of the star, eventually forming a core composed of almost pure helium. This stifles the nuclear reactions in this region, and hydrogen fusion can continue only in a spherical shell surrounding the helium-rich core. The heat generated by this shell may be able to support the core for a while, but as time goes on and more helium is produced, the core's gravity increases. Eventually, the central parts of the star either collapse or undergo some other transformation that can culminate in the fusion of helium-4 into carbon-12. These drastic changes in the core alter both the star's size and the properties of the light it emits. Typically, the star becomes brighter and redder,

2. One exception to this rule is the fusion of helium-3 and helium-4 into a nucleus of beryllium-7, with four protons and three neutrons. However, after beryllium-7 is formed, it undergoes additional reactions, including fusion with another hydrogen nucleus, which eventually produces two helium-4 nuclei. The production of beryllium-7 is therefore just another way to assemble helium-4 nuclei, and does not provide an additional source of energy to the star.

transforming into a red giant. This transition marks the start of a fascinating and a very complex process that ultimately results in the star dying. The star either explodes like a supernova or the nuclear reactions in the core fizzle out, leaving a remnant like a white dwarf behind.

Rather the delve into the gory details of how actual stars transform into red giants, we will here consider a simple model of main sequence stars that illustrates how the steady accumulation of helium can eventually provoke a major change in the structure of the star. Imagine a star composed of two parts, an envelope of hydrogen-rich material surrounding a small helium core. Fusion occurs at the bottom of the envelope, providing heat to support both the star as a whole and the core in its center. It also yields a steady supply of helium that causes the mass of the core to grow over time. As long as the core contains only a small fraction of the total mass of the star, the weight bearing down on the shell and the core does not change much over time. The nuclear reaction rate in the shell therefore remains steady, the temperature of the central part of the star does not change, and its luminosity and surface temperature hold constant. In this state of quasi-equilibrium, the temperature in the core is high enough to support the entire mass of the star against gravitational collapse. Put another way, the energy from the fusion shell keeps the helium nuclei moving around so fast that when they collide with the outer layers of the star, they provide enough of a "kick" to keep particles from drifting inwards. Yet, as the core grows, it must also avoid collapsing under its own gravity. Since a helium atom is four times more massive than a hydrogen atom, it requires significantly more kinetic energy to resist the gravitational pull towards the center of the star. This means that even though the kinetic energy of the particles in the core can support the full mass of the hydrogen-rich star, they can support a helium-rich core with only a small fraction of this mass. Assuming that the hydrogen and helium can be treated like classical gases, the ratio of the maximum possible core mass to the total mass of the star depends on the average mass of the particles in the core and the envelope. For a helium-rich core and a hydrogen-rich envelope, the maximum mass of the core is about 10% of the total mass of the star. This is known as the Schonberg-Chandrasekhar limit, after the scientists who first computed it in 1942.[3] When the mass of the core exceeds this limit, it will begin to collapse, triggering a major reorganization of the star's interior.

3. This is not be confused with the Chandrasekhar limit, which applies to dense objects like white dwarves.

I must emphasize that this is a gross oversimplification of the complex sequence of events that are believed to occur in real stars. In fact, stars with different masses appear to evolve in very different ways. For example, in stars several times more massive than the sun, mixing in the deeper parts of the star prevents a pure helium core from growing steadily from the center of the star, and instead a core forms abruptly as hydrogen is depleted throughout the depths of the star. By contrast, in less massive stars, the core temperature can be low enough and the density high enough that quantum mechanical effects must be taken into account. These complications mean that the Schonberg-Chandrasekhar calculation does not strictly apply to any real star. Astronomers therefore had to develop a variety of methods and tools that allow them to accurately calculate how the internal structure of the star will change as helium accumulates in the core. Remarkably, these more sophisticated calculations also show that main sequence stars begin to become brighter and redder when they have converted about roughly 10% of their mass into helium.

If main sequence stars can survive only until they have fused about 10% of their hydrogen into helium, then we can calculate their life spans from their masses and their luminosities. For example, the mass of the sun is 2×10^{30} kilograms, and its luminosity (or total energy output) is 4×10^{26} watts. Since the sun is close to equilibrium, its luminosity is roughly equal to the power generated by fusion reactions in its core. Einstein's equation $E = mc^2$ tells us that this amount of power would require converting 4 billion (4×10^9) kilograms of mass into other forms of energy every second. The mass of a helium nucleus is about 0.7% lower than the mass of four hydrogen nuclei, so this reduction in mass requires transforming 600 billion (or 6×10^{11}) kilograms of hydrogen into helium every second. In other words, our sun converts 18 quintillion (1.8×10^{19}) kilograms of hydrogen to helium every year. Don't worry, though, because even at this rate the sun would take 11 billion years to convert 10% of its mass to helium and begin its transformation to a red giant. More detailed and careful calculations indicate that the sun can live on the main sequence for about 10 billion years, so this quick back-of-the-envelope calculation is not too far off, and we still do not have to worry about the sun becoming a red giant anytime soon.

Similar calculations allow us to estimate the lifetimes of the other main sequence stars that have well-measured masses and luminosities. What's more, since the mass and the luminosity of main sequence stars are so strongly correlated, we can often infer the life span of such a star from its luminosity alone. For example, say we found a main sequence star 10,000 times as luminous as the sun. From the observed mass-luminosity relation, we know that a star needs to be 10 times as massive as the sun to produce this amount of light. This means that while the star is converting hydrogen into helium at a rate 10,000 times faster than that of the sun, it only has 10 times as much hydrogen available. This star will therefore burn through 10% of its hydrogen 1000 times faster than a solar-mass star, and so can last only 10 *million* years before it becomes a red giant.

This calculation shows that the life span of a main sequence star—like its luminosity—is a very strong function of its mass. The more massive and luminous the star is, the less time it takes for it to become a red giant. While this result does not allow us to know the age of any particular star, it does allow astronomers to estimate the ages of certain collections of stars.

Imagine that a collection of main sequence stars were all formed at the same time. These stars have a range of masses, so if we measured the luminosities and temperatures of these stars just after they formed, we would observe a complete main sequence like the one shown in Figure 10.4. However, it would not take long (astronomically speaking) before the brightest, bluest stars run out of fuel and convert into red giants. In other words, the stars near the top of the main sequence would begin to move off to the right (red) side of the color-magnitude diagram. As time goes on, progressively dimmer and redder stars will also move off the main sequence. If we came back to observe this same group of stars some time later, we could determine how old they are by studying how much of the main sequence remained.

It might at first seem unlikely that such collections of stars would ever occur in nature, but in fact there are several different types of star clusters that appear to contain many stars that formed at a single time in the past. Here we will focus exclusively on a particularly interesting subset of these objects that are known as globular clusters.

SECTION 10.4: THE AGES OF GLOBULAR CLUSTERS

The globular clusters are so named because they appear as fuzzy balls of light when seen through binoculars and small telescopes. More powerful instruments reveal that these objects are actually collections of up to a million stars packed into a region about 100 light-years across: the density of stars here is hundreds of times higher than it is in our neighborhood. Figure 10.7 shows the color-magnitude diagram of the stars from one particular globular cluster. In the lower part of this graph, we can trace a line running diagonally across the plot. This trend is in the right place and has the right orientation to cor-

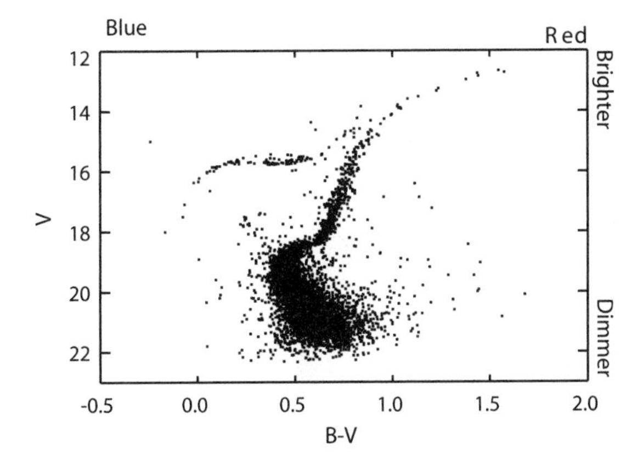

FIGURE 10.7 Color-magnitude diagram of a globular cluster M3. This diagram can be compared with that in Figure 10.4. Note in particular that the diagonal line corresponding to the main sequence near the bottom of the plot appears to be cut off on the blue end. This indicates the cluster has a finite age. The values of *V* on the *y*-axis are larger here than in Figure 10.4 because this plot shows apparent magnitudes instead of absolute magnitudes. (Since all these stars are the same distance away, this just offsets the magnitudes of all of the stars by the same amount.) Data from Ferraro et al. (1997), available at http://vizier.cfa.harvard.edu/viz-bin/VizieR-2?-source=J/A%2bA/320/757.

respond to the main sequence. However, unlike the main sequence formed by nearby stars, this sequence ends abruptly around the middle of the plot, indicating that the brighter, bluer stars of the main sequence are missing from this cluster. Instead, we see an arc of bright red giant stars at the upper right-hand part of the plot, with a horizontal branch extending towards the blue side of the graph.[4]

This distribution of stellar spectral characteristics is consistent with a collection of stars formed at a single point deep in the past. We can therefore attempt to estimate the age of these stars from the extent of the main sequence. In this globular cluster, the brightest, bluest main sequence stars are just slightly redder than our sun. These stars should therefore have luminosities, masses, and life spans similar to those of our sun. Since there are few main

4. Since these bright blue stars fall along a horizontal branch and not the diagonal main sequence, they are not main sequence stars, but are instead stars that have already gone through a red giant phase and are now burning both hydrogen and helium.

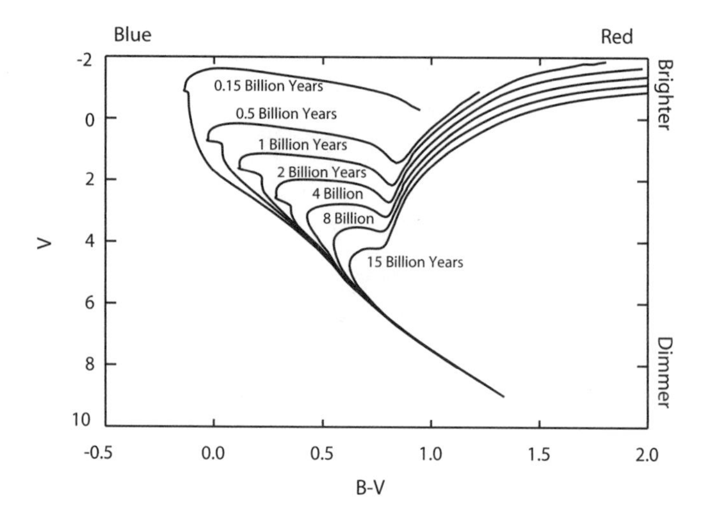

FIGURE 10.8 Schematic color-magnitude diagram of a collection of stars created at the same time, at different times after their creation. This illustrates how these stars progressively transform into red giants. The brightest bluest stars run out of hydrogen first and become bright and red, but as time goes on, redder and dimmer stars leave the main sequence.

sequence stars bluer and brighter than our sun, the cluster must have formed long enough ago that all of these stars have had time to turn into red giants. On the other hand, since the cluster still contains plenty of stars redder and dimmer than our sun, it cannot be too old or else these stars would have also already turned into red giants. In other words, the age of the cluster is *larger* than the lifetimes of all of the bright blue main sequence stars that have already died and *smaller* than the life spans of the faint red stars that are still present. The cluster age is therefore *equal* to the life span of the stars at the blue, bright end of the main sequence—the so-called the main sequence turn-off—which should be just about to transform into red giants. For this particular cluster, those stars are just slightly redder than our sun, so the entire system should be slightly older than the life span of our sun, meaning that its stars formed somewhat more than 10 billion years ago.

This quick analysis gives us only a ballpark estimate of the age of the cluster. To obtain more precise estimates, astronomers compare the spectral data from the cluster to theoretical predictions—like those shown in Figure 10.8—which incorporate far more detailed nuclear physics and hydrodynamics. This series of curves illustrates how stars gradually peel off from the bright blue end of the main sequence to become red giants. We can also gauge the validity of our earlier calculations by noting that the predicted curve of an eight- to fifteen-billion-year-old cluster of stars does in fact bear a close resemblance to the observed data.

Taking a closer look at the data and the theoretical predictions, shown together in Figure 10.9, we can begin to appreciate the challenges involved in this sort of analysis. The data more closely follows the fifteen-billion-year-old curve than either the ten- or twenty-billion-year curves. However, the spread in the observed data is sufficiently large that we cannot just look at the data and estimate the age of the cluster to within a billion years. For this reason, astronomers have to use careful statistical analyses to obtain a precise measure of age from these sorts of data.

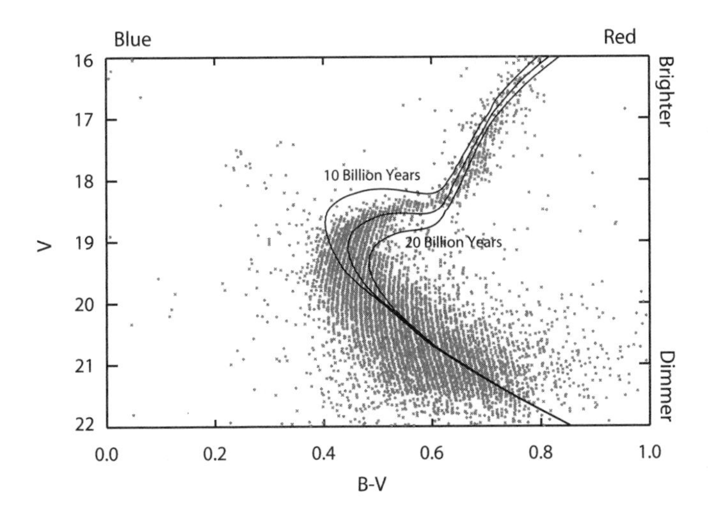

FIGURE 10.9 A closer look at the main sequence part of M3 (illustrated in Figure 10.7) with the theoretical curves for ten-, fifteen-, and twenty-billion-year-old clusters. The data and the curves are aligned by eye for purposes of illustration. In practice, globular clusters are sufficiently far away that the parallax method of measuring distance does not work, so astronomers use color information and the position of the horizontal branch to convert observed apparent magnitudes into absolute magnitudes and to align data and theory. By comparing the shape and location of the main sequence turn-off of M3 with the curves, we can see the cluster is closer to fifteen billion years old than ten or twenty billion years old. With a more careful analysis that accounts for the scatter in the data etc., astronomers have estimated the age of the cluster as thirteen billion years, give or take a billion years or so.

Of course, even with a good measurement of the shape of the main sequence turn-off, the accuracy of the derived age still depends on the reliability and applicability of the theoretical model. The shapes of the theoretical curves depend somewhat upon the composition (helium-to-hydrogen ratio, etc.) of the stars in the cluster, and certain clusters might even contain multiple populations of stars, each with a slightly different main sequence turn-off. These factors must be taken into account when comparing the data and theory to obtain a reasonable age estimate for any given cluster. Occasionally, these model calculations require revisions that can alter the overall age estimates of globular clusters. For example, in 2004 nuclear physicists announced the results of new investigations into a reaction where a nitrogen-14 nucleus and a proton fuse together to form oxygen-15. These new data indicate that this process occurs less readily than was previously thought. This discovery impacts the age estimates of globular clusters because this reaction is an important part of a cycle that can facilitate the fusion of hydrogen into helium, provided there are enough carbon, nitrogen, and oxygen available. Updating the models with these new data alters when and how quickly the stars move off of the main sequence; if these new findings are correct, the age estimates of the globular clusters need to be increased by about a billion years.

Clearly there is still work to be done before the age of distant stars can be measured with the same precision and accuracy as the age of our solar system. Yet the basic idea behind this method of measuring stellar ages is reasonable, and astronomers have even found ways to independently confirm the results of these analyses. For example, the characteristics of faint stars called white dwarfs in clusters have been used to place constraints on their ages. Astronomers can sometimes even detect small amounts of radioactive nuclei in these distant stars and apply a form of radiometric dating to these systems. These methods all agree that the oldest stars in the oldest globular clusters are about twelve or thirteen billion years old, give or take a billion years. Even with ten million centuries of uncertainty, these dates are very interesting because they suggest that globular clusters are very ancient objects and that some stars had been shining for billions of years before our solar system formed. In fact, the stars in these globular clusters may be nearly as old as the universe itself.

SECTION 10.5: FURTHER READING

For a good general introduction to astronomy, see Roger A. Freedman and William J. Kaufmann *Universe,* 6th ed. (Freeman and Co., 2001). For more detailed works on stellar astronomy, see R. J. Taylor *The Stars: Their Structure and Evolution* (Cambridge University Press, 1994) and R. Kippenhahn and A. Weigart *Stellar Structure and Evolution* (Springer-Verlag, 1994).

A good resource for finding raw data about different types of stars can be found at http://vizier.cfa.harvard.edu/vizier, and many technical articles can be found at www.arxiv.org.

A detailed discussion of globular clusters can be found in K. M. Ashman and S. E. Zepf *Globular Cluster Systems* (Cambridge University Press, 1998). A good review of the issues involved in dating globular clusters is B. W. Carney and W. E. Harris *Star Clusters* (Springer, 2000).

Interesting articles on main sequence turn-off dates are B. Chaboyer "The Age of the Universe" *Physics Reports* 307 (1998): 23–30, R. Gratton et al. "Age of Globular Clusters in Light of Hipparcos: Resolving the Age Problem?" *Astrophysical Journal* 494 (1998): 96–110, and R. Jiminez "Towards an Accurate Determination of the Age of the Universe" in *Dark Matter in Astrophysics and Particle Physics*, ed. H. V. Klapdor-Kleingrothaus and L. Baudis (Institute of Physics Publishing, 1999). For more details about the recent revision of the main sequence turn-off ages, see G. Imbriani et al. "The Bottleneck of CNO Burning and the Age of Globular Clusters" *Astronomy and Astrophysics* 420 (2004): 625–629 and R. C. Runkle et al. "Direct Measurement of the $^{14}N(p, \gamma)^{15}O$ *S*-Factor" *Physical Review Letters* 94 (2005): 082503, on-line at www.arxiv.org/abs/nucl-ex/0408018.

For other methods of measuring the age of globular clusters, try Brad Hansen et al. "White Dwarf Cooling Sequence of the Globular Cluster Messier 4" *Astrophysical Journal* 574, no. 2 (2002): L155–L158, on-line at www.arxiv.org/abs/astro-ph/0205087, and J. Truran et al. "Probing the Neutron-Capture Nucleosynthesis History of Galactic Matter" *Publications of the Astronomical Society of the Pacific* 114 (2002): 1293–1308.

CHAPTER ELEVEN

Distances, Redshifts, and the Age of the Universe

For a very long time, scientists and other scholars have been puzzling over how and when the universe came into existence. During the last century, increasingly powerful telescopes have shown us that the universe is a vast place, with dust, gas, galaxies, and more enigmatic material strewn over billions of light-years. These observations have also revealed that the universe we see today is probably not infinitely old, but instead arose from a singular and formative event—called the Big Bang—at a definite, measurable time in the distant past. While the basic concept of the Big Bang has been around for about a hundred years, robust and precise constraints on the timing of this important event have appeared only in the last decade.

Some of the most valuable data we currently have about the age and history of the universe derives from the colors and distributions of galaxies. These clumps of stars and gas shine with the combined light of billions of stars and can be seen over distances stretching billions of light-years, so they can provide information about the conditions of the universe in many different places and at many different times. These data not only can help us pinpoint when the Big Bang happened, but also clarify the nature of this event, which is quite difficult to describe and conceptualize in everyday terms.

SECTION 11.1: OBTAINING REDSHIFTS

For cosmologists, one of the more illuminating characteristics of galaxies is a quality known as the redshift. The redshifts of galaxies are derived from their spectra, which often show narrow features like those found in Figure 11.1.

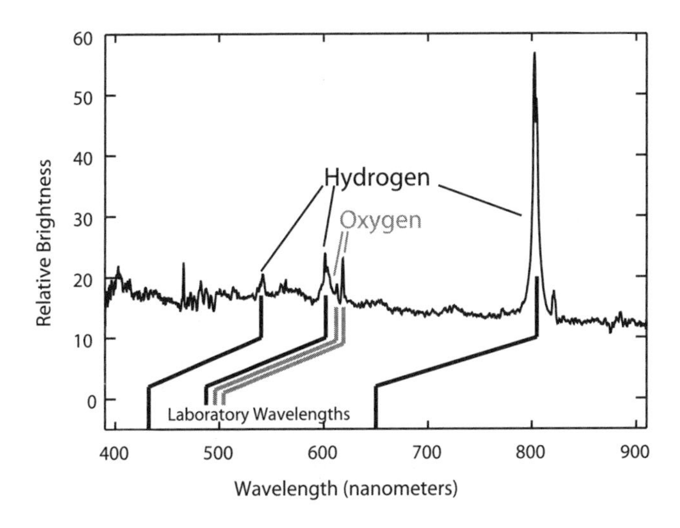

FIGURE 11.1 Spectra of a particular galaxy measured by the Sloan Digital Sky Survey (specID: 53140-1625-454, data available from www.sdss.org). This plot shows the amount of light emitted by the galaxy at different wavelengths. These data show a number of narrow spikes that correspond to the discrete transitions of various atoms. Spikes due to hydrogen and oxygen are identified and marked. The observed wavelengths of these features are all shifted to longer wavelengths compared to the wavelengths of such features observed in the laboratory (indicated by the lines near the bottom of the plot).

These lines are typical of radiation produced by atoms and molecules. Each type of atom strongly absorbs or emits light at only certain discrete wavelengths. For example, sodium atoms can generate yellow light with a wavelength of 589 nanometers, while mercury atoms can produce bluish light at 436 nanometers. Such characteristic wavelengths reflect subatomic processes that are best described using the equations of quantum mechanics. Put simply, these equations indicate that the electrons surrounding a given nucleus can exist only in a finite number of configurations for any significant length of time. When the electrons change from one of these configurations to another, the atom must either take in or give off some radiation to conserve energy, momentum, and angular momentum. This transition radiation has a well-defined wavelength that is determined by the initial and final states of that particular atom's electrons. Different transitions produce light at different wavelengths, generating a pattern of lines in the spectrum specific to each type of atom in the periodic table.

The patterns of lines generated by different atoms have been measured in the laboratory with great precision, which allows us to identify the various elements present in a particular galaxy. For example, the spectrum of the galaxy illustrated in Figure 11.1 has three spikes at 800 nm, 600 nm, and 530 nm. These three numbers are part of a sequence characteristic of hydrogen, the most common element in the universe: (2/1) × 400 nm, (3/2) × 400 nm, (4/3) × 400 nm. However, while the *pattern* of spikes is consistent with hydrogen, the *positions* of the features are not. In the laboratory, these lines occur at about (2/1) × 328 nm, (3/2) × 328 nm, and (4/3) × 328 nm, which means the features in the galaxy occur at wavelengths 23% longer than they do here on earth.

Does this mean that the features are due to some material other than hydrogen? Not likely, since there are other features in the spectrum that correspond to other elements. The two spikes around 620 nm, for instance, have a separation consistent with two lines in the spectrum of oxygen, but in the lab they occur around 500 nm. Again the features in the galaxy differ from the features measured on earth by a factor of 23%. With patterns ascribed to several different elements, it is difficult to argue that we have misidentified materials in the galaxy. Instead, it appears that the wavelength of all the light from the galaxy has increased by 23% between when it was produced in the galaxy and when our equipment observed it.

The spectra from almost all of the galaxies that have been studied show similar shifts to longer wavelengths. These are known as redshifts because longer wavelengths correspond to the red end of the visible spectrum, and they are quantified by the fractional change in the wavelength of the light. For the above galaxy the redshift is 0.23, and other galaxies have redshifts that range from a few percent up to a factor of 7.[1] The observation of galactic redshifts has important implications for the dynamics and history of the universe. However, before we rush to interpret or explain these measurements, we should first explore how the redshift of a galaxy depends on its position in space and its distance from us.

SECTION 11.2: MEASURING DISTANCES

Unlike the redshift, which can be derived relatively easily from a good spectrum, the distance to a galaxy is a very challenging thing to measure. As we discussed in the previous chapter, there is a way to directly measure astro-

1. Only a very few, very nearby galaxies have measurable shifts to shorter wavelengths.

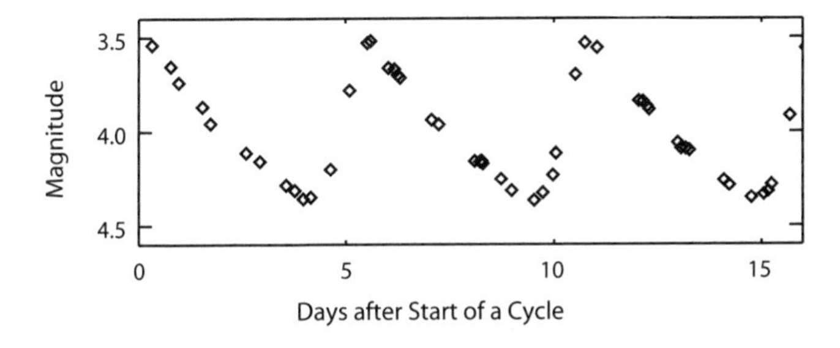

FIGURE 11.2 The brightness variations of Delta Cephei, a typical cepheid star. A cepheid brightens quickly and then dims more slowly in a repeating cycle. For this cepheid, the cycle takes 5.3 days to complete. Other cepheids can take up to a month to complete one brightness cycle. However, all cepheids share this saw-tooth pattern in their variations. (Data from T. J. Moffett and T. G. Barnes "Observational Studies of Cepheids II: BVRI Photometry of 112 Cepheids" *Astrophysical Journal Supplement Series* 55 (1984): 389–432.)

nomical distances based on simple trigonometry. Unfortunately, this method cannot be applied to objects far outside our own galaxy because the apparent motion of the objects is too small to be measured. Other, more indirect techniques are therefore required to estimate the vast distances separating galaxies.

Many methods for measuring the distances to galaxies are based on the apparent brightness of certain astronomical objects. As we saw in the last chapter, the apparent brightness of an object decreases with distance, so an object located 200 meters away appears four times fainter than a similar object located 100 meters away. Turning this around, if we know how much light the object generates (its luminosity), then we can use the apparent brightness of the object to estimate its so-called luminosity distance away from us. At the present moment, there is no way to accurately calculate the luminosity of any extragalactic astronomical object from theory alone. However, there are certain types of objects where the luminosity can be estimated based on other characteristics of the observed light.

A good example of this sort of object is a cepheid, a type of star whose luminosity varies with time in a characteristic cycle (shown in Figure 11.2). Cepheids first rapidly increase in brightness, then more slowly dim until the brightness reaches its original level and the cycle begins again. This cycle can take days or weeks to complete, depending on the cepheid. Cepheids are

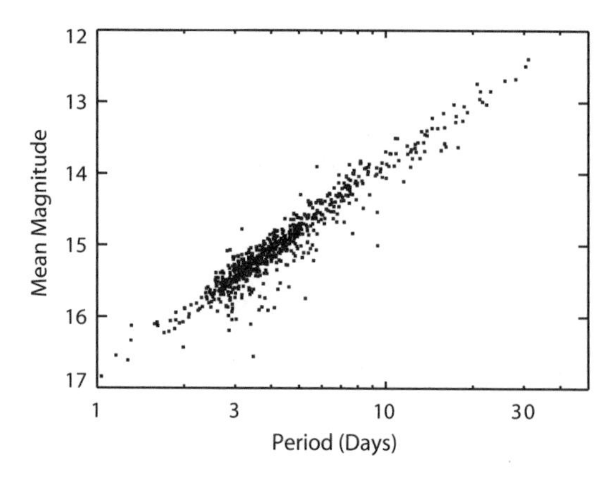

FIGURE 11.3 The relationship between the mean brightness and the period of brightness variations for cepheids in the Large Magellanic Cloud (based on data in A. Udalski et al. *Acta Astronomica* 49 (1999): 223). Since all of these cepheids are roughly the same distance away from us, the brightness variations reflect real variations in the luminosity of the cepheids. The brightness of these objects is correlated with the period, so by measuring how fast a cepheid changes its brightness, we can obtain a reasonable estimate of its luminosity.

thousands of times more luminous than the sun, so they can be seen from very far away, even in other galaxies. Since cepheids are a particular, identifiable type of star, they would be good distance indicators if we could determine how luminous they are. Fortunately, there is a natural laboratory for understanding the characteristics of cepheids: the Magellanic Clouds.

The Large and Small Magellanic Clouds are two satellite galaxies that orbit around the Milky Way. The stars in each cloud are roughly the same distance from us, so if all cepheids had the same luminosity, the cepheids in the Large (or Small) Magellanic Cloud would all have the same apparent brightness. In reality, as shown in Figure 11.3, the brightness of the cepheids in the Large Magellanic Cloud ranges over five magnitudes. The luminosity of these objects can therefore vary by about a factor of 100.

Even though the cepheids have a wide range of luminosities, they can still be used to estimate distances because the brightness of a cepheid is correlated with its period—the time it takes to go through one brightness cycle. This means that a cepheid that is more luminous also takes longer to brighten or dim, so we can estimate a cepheid's luminosity from its period and also infer how far away it is. For example, say we find a cepheid in another galaxy with a period of about 10 days and a mean magnitude of 25. This cepheid is ten magnitudes, or 10,000 times, as faint as a cepheid with a comparable period in the Large Magellanic Cloud. Therefore, that cepheid—and the galaxy that contains it—must be 100 times as far away as the Large Magellanic Cloud.

If we want to know how far away this galaxy is in light-years, then we have to know the distance to the Large Magellanic Cloud. Fortunately, there are cepheids in our own galaxy, and some of these are close enough that astronomers can use the parallax method to obtain accurate distance measurements, thereby determining their luminosity. Relating these measurements to the cepheids in the Magellanic Clouds tells us that the Large Magellanic Cloud is about 150 thousand light-years away (other methods of measuring the distance to this object have yielded roughly similar results). The galaxy in the above example must therefore be about 15 million light-years away.

While cepheids are quite bright compared to the sun, they can be identified only in galaxies within about 100 million light-years of us. This may seem like a very great distance, but the vast majority of galaxies are farther away than this. Much brighter objects are therefore needed to estimate the distances to these more remote objects. Recently Type Ia supernovae have emerged as a powerful tool for measuring the great distances to such far-flung galaxies.

Supernovae are catastrophic events that can produce as much light as a billion suns for a couple of weeks. These events are most likely powerful explosions that occur when the nuclear reactions in the core of a star can no longer prevent it from collapsing under its own gravity. This can happen through a number of different processes, and indeed there are several distinct types of supernovae, each of which has particular identifiable features in their spectra. Type Ia supernovae—among other things—lack the sequence of lines characteristic of hydrogen, so these explosions likely involve hydrogen-poor stars. Specifically, they probably occur in former white dwarfs, the burnt-out cores of low-mass main sequence stars. Nuclear reactions have largely stopped in these objects, but when another nearby star dumps some additional mass onto the dwarf, it can be knocked out of equilibrium and explode.

While such supernovae are relatively rare—occurring perhaps once a century in any galaxy—a number of these events have been observed in galaxies with cepheids or other distance indicators, enabling astronomers to estimate the total amount of light they generated. These data show that the peak luminosity of these events can vary by about a factor of three. However, more luminous supernovae tend to fade away more slowly, so just as with the ce-

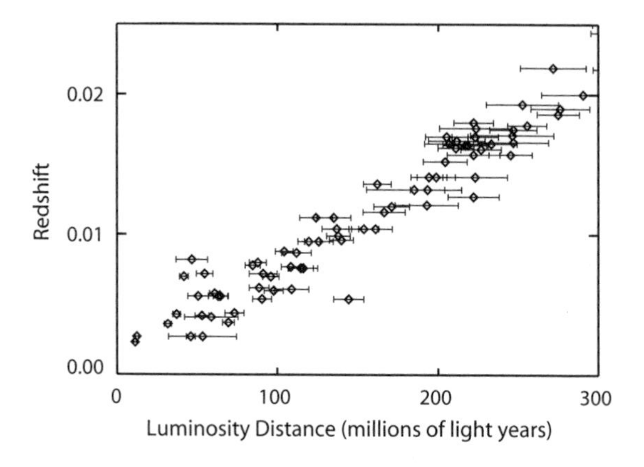

FIGURE 11.4 A Hubble diagram showing the redshifts of different galaxies versus the luminosity distance deduced from the brightnesses of Type Ia supernova (based on data in John L. Tonry et al. "Cosmological Results from High-z Supernovae" *Astrophysical Journal* 594 (2003): 1–24). Each point represents a single galaxy, and the horizontal error bars show the uncertainty in the distance measurement (uncertainties in the redshifts are much smaller). Note that the farther away the galaxy is, the larger its redshift.

pheids, the way the light from this particular type of supernova changes with time reveals how much light the explosion generated. Therefore, we can again compare the total energy output of this event to its observed brightness and calculate how far away the supernova and its host galaxy are.

With both redshift and distance data for multiple galaxies, we can make what is known as a Hubble diagram, a graph of the redshift versus distance (Figure 11.4). Such diagrams show quite clearly that the farther away the galaxy is, the larger its redshift. This correlation between redshift and distance must reflect some fundamental characteristic of the dynamics and history of the universe. However, there is more than one way to interpret a Hubble diagram, so we need additional astronomical and cosmological observations to evaluate these alternatives and to find the most likely explanation of these data.

SECTION 11.3: A WRONG WAY TO LOOK AT HUBBLE DIAGRAMS

One interpretation of the Hubble diagram—which is so seductively intuitive that it often finds its way into popular introductions to cosmology—is that the redshifts are Doppler shifts due to the relative motions of galaxies. This is an appealing explanation of galactic redshifts because Doppler shifts can be easily studied and demonstrated on earth. For example, Doppler shifts in sound waves are what cause the pitch of an ambulance's siren to change abruptly as it passes us on the street. Similar shifts in the wavelength of electromagnetic radiation from moving objects are usually too subtle for us to see with our own eyes, but they can be detected with Doppler radar systems, GPS receivers, and so on.

If we attempt to interpret galactic redshifts as Doppler shifts, then the fact that the light has been shifted to longer wavelengths means that the galaxies must be moving away from us (just like the drop in the pitch of a siren implies that the vehicle has passed by). Furthermore, we can calculate how fast the galaxies are moving from the size of the redshift. If the redshift is much less than one, then the redshift is almost exactly the speed of the galaxy relative to the speed of light, so a galaxy with a redshift of 0.1 is traveling away from us at about one-tenth the speed of light. For higher speeds and larger redshifts the calculation is a little more complicated—a galaxy with a redshift of 1.0 is moving at about six-tenths the speed of light—but really only a little additional algebra is needed to make this estimate.

Taking the redshifts as a measure of galactic motion, then the Hubble diagram indicates that all galaxies are moving away from us, and that galaxies farther away from us are moving faster. This odd situation is illustrated here:

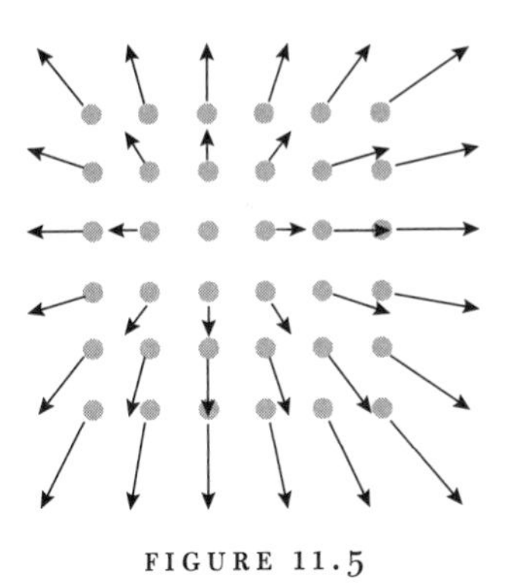

FIGURE 11.5

The gray dots represent galaxies. The galaxy with no arrow is the one we are living in and the arrows on the other galaxies indicate how quickly and in what direction they appear to be moving.

This cartoon implies that we are so cosmically unpopular that all galaxies in the universe are trying to get away from us, but this need not be the case.

Imagine instead all the galaxies flying away from a special point in space, with a speed proportional to their distance from that point, as shown here:

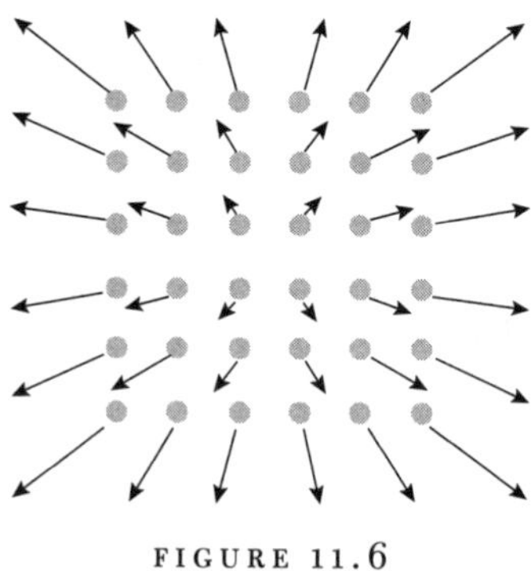

FIGURE 11.6

In this case, anyone in any galaxy in this group would observe the same simple relationship between distance and relative velocity as we find in the Hubble diagram.

A universe filled with galaxies flying away from some point is, unfortunately, close to what many of us envision upon hearing the words "Big Bang." However, this is not what cosmologists mean when they talk about this event, and this scenario does not mesh with other observations of the large-scale structure of the universe. If all the galaxies were launched from a special point in space, then we would expect that the characteristics and distribution of the galaxies would depend on their distance from this central point. For example, we might predict that galaxies further from this point would have lower masses than galaxies closer in (or be a different type, or have a different chemical composition). We might also guess that there would be more galaxies nearer to this point than further away or vice versa. So far, no observation of the large-scale structure of the universe shows any pattern that would support such ideas.

Consider Figure 11.7, which shows the distribution of galaxies in space as measured by the Sloan Digital Sky Survey. These data show no large-scale structures that clearly point to a particular location in the universe from which all the galaxies could have originated. Indeed, these and all other observations to date indicate that the universe is basically homogeneous on large scales, with the same types of galaxies in roughly similar distributions throughout all of space. There are definitely clusters and groups of galaxies in various places, but these structures do not extend to fill a large fraction of the observable universe. This contradicts the notion that there could be one specific point in the universe where all the matter could come from.

The large-scale homogeneity in the structure of the universe indicates that the patterns in the apparent motions of galaxies are due to a broader phenomenon, which affects all regions of space equally. Strictly speaking, this does not rule out a model of the universe with galaxies moving through space. For example, we could imagine an infinite universe with all galaxies gradually moving further and further apart, in which case there need not be any special central location in the universe. Such "Newtonian" cosmologies are often useful for instructional and computational purposes. However, even these models yield a rather misleading picture of how most cosmologists think the universe works. In fact, the best explanation available today for the data in the Hubble diagram is far stranger than any cosmic explosion, but it is something that occurs naturally in the context of general relativity.

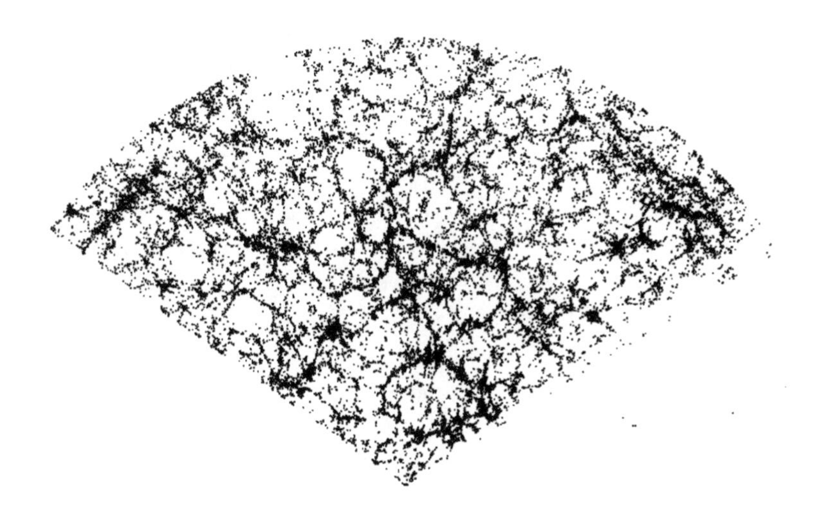

FIGURE 11.7 A sample of the Sloan Digital Sky Survey, showing the distribution of tens of thousands of galaxies in a wedge of space. The tip of the wedge marks the location of the earth, and the most distant galaxies are roughly two billion light years away. The distribution of galaxies is on average the same at every part of this wedge, and there are no gross patterns that would suggest that galaxies are moving out from some special point in space. (Data obtained from the spectro pipeline available at spectro.princeton.edu.)

SECTION 11.4: A BRIEF INTRODUCTION TO GENERAL RELATIVITY

General relativity is the theory that currently provides the most accurate method for calculating how objects move under the influence of gravity. While the mathematical manipulations required to extract predictions from this theory are quite intimidating, the basic premise behind these computations is easy to express, if perhaps difficult to accept: gravity is not a force so much as a distortion in the geometry of space and time.

Forces are an important part of the classical mechanics of Newton and Galileo, which states that in the absence of outside forces, an object will move at a constant speed in a straight line. Outside forces are therefore responsible for any changes in the speed or direction of an object's movement. However, before we can calculate how an object will respond to a given force, we often need some additional information about the characteristics of the object. For example, we usually need to know the mass of an object before we can determine how fast it will accelerate in response to a particular force. In many cases, different objects can even experience different forces when placed in the same environment. For example, imagine we had a proton, electron, and neutron all sitting next to a large positive charge. In this situation, the proton is repelled from the charge, the electron is attracted to it, and the neutron feels no force at all.

Gravity is unique in that all objects subjected to a fixed gravitational field move in the exact same way. A famous demonstration of this involves a feather and a hammer. Both objects are held above the surface of the earth in a vacuum. When they are released, both objects fall at the same rate and hit the ground at the exact same time. In fact, we could have used any object, from a grain of sand to a dump truck, from a piece of gold to a clod of dirt. So long as we remove air resistance, all these objects will move in the exact same manner.

In classical physics, this startling result is essentially a coincidence. Say the hammer had a mass a thousand times larger than that of the feather. The gravitational force on the hammer is then a thousand times stronger than the force on the feather. However, the hammer also takes a thousand times as long to reach a given speed in response to a given force, so the two objects accelerate at the same rate and move in the same way.

By contrast, in general relativity the fact that all objects respond in the same way to a gravitational field reveals a fundamental aspect of how gravity really works. The only other situation where the motion of objects does not depend on their intrinsic qualities is when *no* outside force is operating, and in that case any object moves at a constant speed in a straight line. A straight line, of course, is the shortest distance between two points in Euclidean geometry, so the path that objects take in the absence of outside forces has a specific geometrical definition. General relativity posits that objects in a gravitational field follow a similar geometrically defined path. In essence, we are asked to imagine that the gravitational fields produced by massive objects correspond to distortions in the geometry of the space, which cause the "shortest distance" between any two points to no longer resemble a straight line. Any object will then follow this curving trajectory unless other, nongravitational forces are operating. General relativity therefore suggests that gravity is not a force that causes particles to deviate from a straight-line path; it is a change in the definition of a straight-line path itself.

As a side note, I must point out that the relevant displacement in general relativity actually involves changes in both space and time. Imagine we had two spaceships moving between two points in space. One spaceship follows the typical curved orbit around earth and the other uses its thrusters to travel along a straight line. If we used a ruler or the odometers on the spacecraft to measure the distances they traveled, we would *not* find that the spacecraft that took the curved path traversed a smaller number of kilometers. However, after completing the journey, the two spacecraft would have different readings not only on their odometers, but also on their internal clocks, and one particular combination of these temporal displacements and spatial path-lengths is at an extreme for the spacecraft in orbit.

General relativity is clearly an intriguing way to look at gravity. However, general relativity is not just theoretical speculation motivated by elegant mathematical formulations. It explained phenomena—like certain quirks in the orbit of Mercury—that could not be accounted for using classical gravity. It also predicted effects that were later confirmed by observations, such as the bending of starlight by the sun. General relativity therefore provides the best available model for understanding how gravity works, both theoretically and experimentally. It can also explain the relationship between redshift and distance observed in the Hubble diagram.

SECTION 11.5: THE EXPANDING UNIVERSE

General relativity includes a novel process that can alter the wavelength of light from distant galaxies: an expanding universe with a changing scale factor. Say we have a universe filled with evenly spaced galaxies (imagine the pattern continuing infinitely in every direction and in three dimensions):

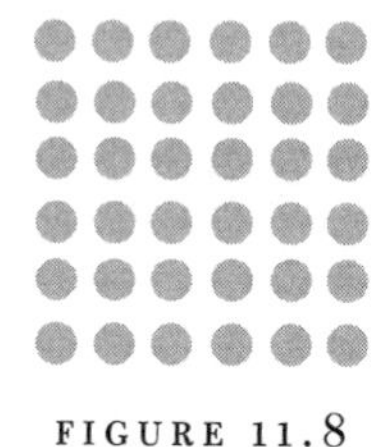

FIGURE 11.8

The distances between galaxies are obviously tied to the geometry of the universe, and according to general relativity, this geometry is affected by the amount of matter in the universe. This means the distances between galaxies can change. We will explore these changes in more detail in the next chapter, but for now, let us consider a simple case where after some amount of time the spacing between the galaxies has increased by a factor of two, as shown here:

FIGURE 11.9

This change in the spacing between galaxies did not happen because the galaxies themselves moved through space, as they did in the "explosive" model discussed previously. Instead, the amount of "space" between the galaxies has increased due to a change in a fundamental geometrical parameter, which is called the scale factor of the universe. In this example, the scale factor doubled between the two images.

Doubling the scale factor not only doubles the distance between any pair of galaxies, it also doubles the distance between the crests of any electromagnetic wave, and so doubles the wavelength of any light propagating through space. A changing scale factor therefore provides a mechanism for generating redshifts in the light from distant galaxies. In fact, at the present moment, this is the best explanation we have for the observed redshifts, because it is consistent not only with these data but also with a host of additional cosmological observations (some of which are discussed in the next chapter). It is certainly a better model than the very simplistic explosive scenario described above. If nothing else, it can accommodate the observations that suggest that the universe is basically homogeneous on large scales. An increasing scale factor operates throughout the entire universe and therefore does not require a special central location.

Since the light from distant galaxies has shifted to longer wavelengths, the scale factor of the universe must have gotten larger since this light was generated. A universe with an increasing scale factor is often described as an "expanding universe"; however, one should not read too much into this phrase. For example, just because the universe is expanding does not mean it is spreading out from some location into some larger space outside the universe. For all we know, the universe could be infinitely large, and changes in the scale factor do not change the size of the universe (two times infinity is still infinity). While it is possible that the universe is finite and forms a well-defined compact object in some higher-dimensional space, we have no observational data that clearly support this idea or quantify the characteristics of that higher-dimensional space, so it is probably best to leave such speculations aside for now.

Also, note that even though the universe is expanding, not everything in the universe is getting bigger. Neither you, me, this book, the earth, nor even the entire galaxy are getting any larger due to the expansion of the universe. Recall that general relativity posits that massive objects distort the geometry of space and time and alter the path that objects naturally take in the absence of other forces. When two objects are held together by another force, like electromagnetism, the distance between the objects is unaffected by this geometrical distortion, except in extreme situations where the geometry changes by a large amount over the separation of the two objects (which is likely to happen only if they fall into a black hole). This means that the typical distance between electrons in an atom or between atoms in a solid object is unaffected by the expansion of the universe. Even galaxies, which are held together by gravity, do not expand with the universe because the local geometrical distortions produced within the galaxy are strong enough to prevent the stars and other objects within it from dispersing. The expansion of the universe can affect the distance between isolated galaxies only because these objects are not bound together and can therefore follow the "straight" paths defined by the geometry of the universe. Similarly, universal expansion influences the wavelength of freely propagating light because no force constrains the distance between the crests of a particular electromagnetic wave.

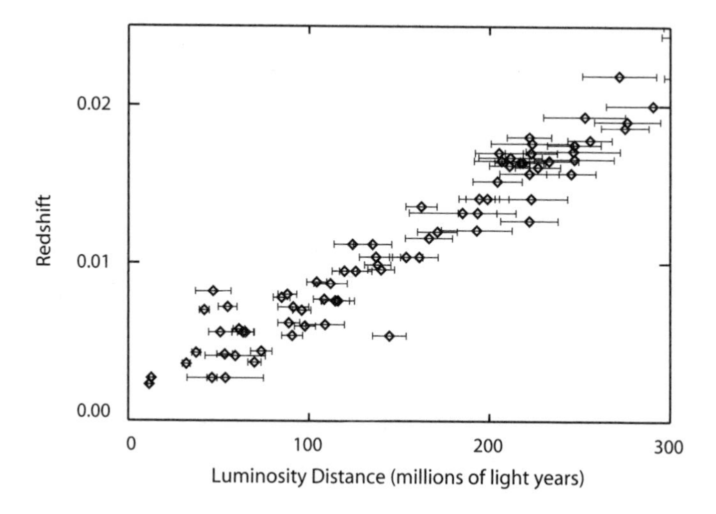

FIGURE 11.10 A Hubble diagram showing the redshifts of different galaxies versus their luminosity distance deduced from the brightness of Type Ia Supernovae.

SECTION 11.6: REINTERPRETING THE HUBBLE DIAGRAM AS A HISTORICAL DOCUMENT

In the context of an expanding universe, redshifts do not reflect the specific characteristics of the source galaxies; instead they indicate how much the entire universe has expanded during the time they were in transit. This means that the redshifts provide a record of the expansion history of the universe. To see how we can extract this record, let us take another look at the Hubble diagram (Figure 11.10).[2] This plot indicates that a galaxy roughly 150 million light-years away has a redshift of 0.01, so the wavelength of light from the galaxy has increased by one percent during its journey to us, and the scale factor of the universe has increased by one percent during this time. The scale factor when the galaxy emitted the radiation was therefore one percent smaller than it is today. If we arbitrarily set the scale factor of the universe equal to 1 today, then the scale factor was 0.99 when the galaxy emitted the light. Following a similar procedure, we can calculate the scale factor of the universe at the time when the light we see today was emitted from every single galaxy in the Hubble diagram, and construct the plot in Figure 11.11.

2. Based on data in John L. Tonry et al. "Cosmological Results from High-z Supernovae" *Astrophysical Journal* 594 (2003): 1–24.

This graph (which basically the previous graph flipped upside-down) shows that as the distance to the galaxy increases, the scale factor when the galaxy produced the light decreases. Light from more distant galaxies takes longer to reach us, so this implies that the scale factor of the universe has been increasing with time. Indeed, we can use the inferred distance to compute the amount of time the light has been traveling and make a plot showing how the scale factor of the universe has changed with time.

In principle, it is a fairly simple task to convert the distance to a galaxy into an estimate of the time the light took to reach us. Light travels at a well-defined speed (about 300,000 kilometers per second), and a light-year is defined as the distance light travels in a year, so a galaxy 150 million light-years away should have emitted the light we receive today 150 million years ago. Unfortunately, the situation becomes more complicated in an expanding universe, because the distance between the galaxies was changing while the light was in transit. These complications are not very important for the Hubble diagram below, because the scale factor only changes by a few percent during the time the light travels. However, recent supernova search programs have managed to find a significant number of supernovae with redshifts around and above 1, corresponding to a change of a factor of 2 in the wavelength of the light and the scale factor of the universe. For these very distant objects, the effects of the changing scale factor can no longer be ignored.

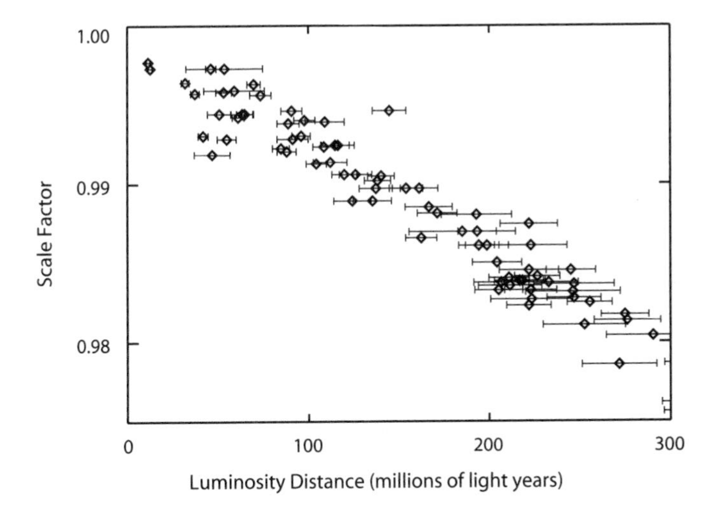

FIGURE 11.11 The scale factor versus luminosity distance for nearby galaxies/supernovae.

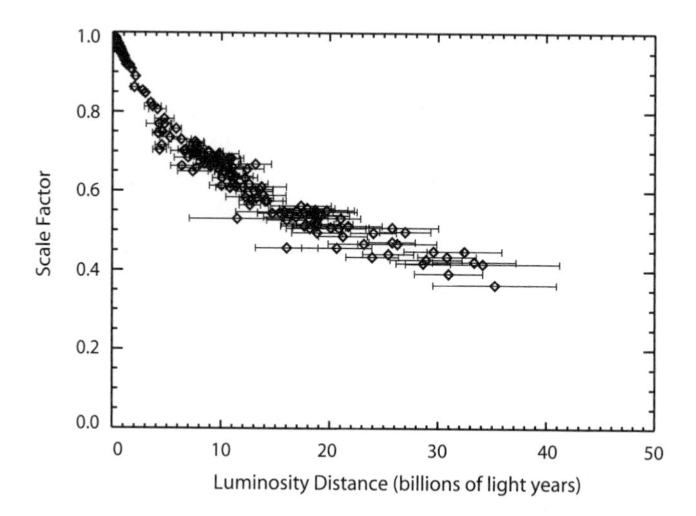

FIGURE 11.12 The scale factor versus luminosity distance from more distant supernovae.

Some of these new observations are plotted in Figure 11.12.[3] The apparent distance of the furthest galaxies is over 30 billion light years. However, the light from these objects has not been traveling through space for over 30 billion years. Instead, these distance estimates are highly inflated due to the large changes in the scale factor. Fortunately, we can account for this and deduce the time it really took the light to travel between the galaxy and us, with some assumptions and approximations.

First of all, remember that the luminosity distance is derived from the apparent brightness of the supernovae. Classically, there is a simple relationship between brightness and distance: an object twice as far away appears four times dimmer. However, in an expanding universe things get more complicated. The expansion of the universe forces the light to become spread over a progressively larger volume and thus causes objects to appear dimmer than we would otherwise expect. Furthermore, as the wavelengths of the individual photons are stretched, the energy per photon decreases, further reducing the apparent brightness of the source. If we assume that Euclidean geometry applies on these large scales (an assumption we will return to in

3. Compiled in A. Riess et al. "Type Ia Supernova Discoveries at $z > 1$ from the Hubble Space Telescope: Evidence for Past Deceleration and Constraints on Dark Energy Evolution" *Astrophysical Journal* 607 (2004): 665–687, and A. Riess et al. "New Hubble Space Telescope Discoveries of Type Ia Supernovae at $z > 1$" (2006), on-line at www.arxiv.org/abs/astro-ph/0611572.

the next chapter), we can correct for the expansion-induced dimming of the galactic light and create a graph of the scale-factor as a function of the distance between the galaxy and us today, which is known as the coordinate distance (Figure 11.13).

The longest coordinate distances are about fifteen billion light-years, well less than the largest luminosity distances. However, this still does not mean that the light from these galaxies has been in transit for fifteen billion years. The coordinate distance is the distance between galaxies *today*; since the universe has expanded with time, the galaxies were closer together when the light began its journey, and the total distance the light traveled is somewhat less than the coordinate distance. Correcting for this effect requires a little more math, but afterwards we finally have the desired graph of the scale factor versus time (Figure 11.14). This graph clearly shows that as we go further into the past, the scale factor gets progressively smaller. Indeed, the data are more or less following a straight line, showing that the scale factor has increased with time at a fairly constant rate for the last ten billion years. Extrapolating this trend further back in time, we find the scale factor would be zero about fifteen billion years ago. A scale factor of zero corresponds to zero distance between adjacent galaxies, so the universe was infinitely dense at this time. This singularity in the density of the universe is the real Big Bang, a time when all of our models of the universe break down, leaving us with no way to

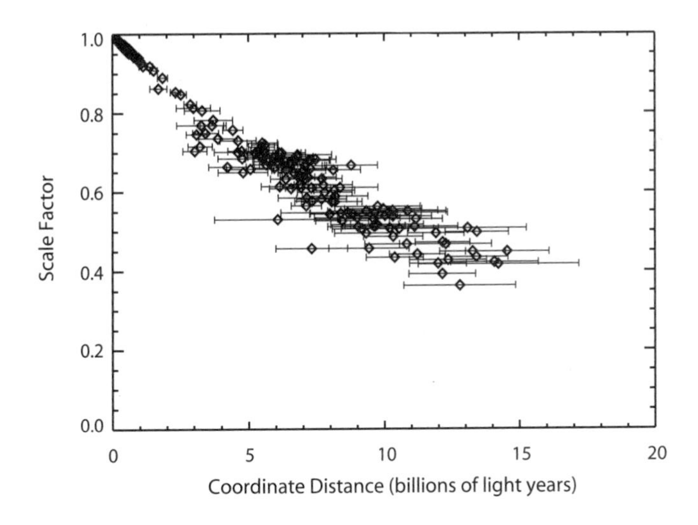

FIGURE 11.13 The scale factor versus coordinate distance (assuming the universe has euclidean spatial geometry).

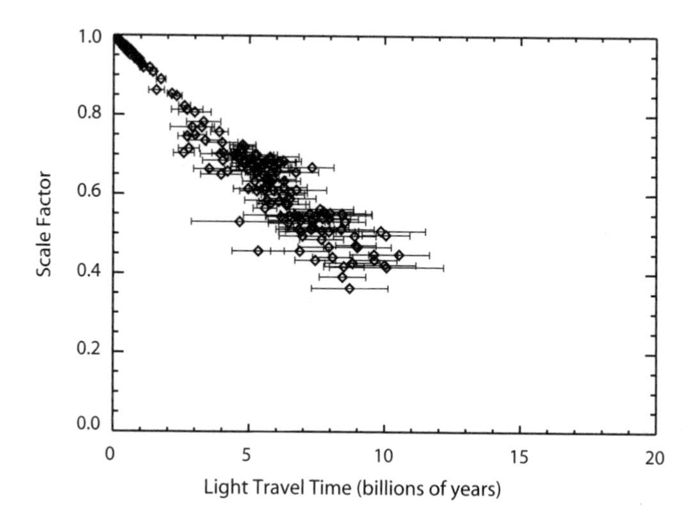

FIGURE 11.14 The scale factor versus light travel time (assuming the universe has euclidean spatial geometry).

guess what was happening at or before the moment. The timing of such a singular formative event is certainly a good estimate of the age of the universe.

A simple straight-line extrapolation of the data can provide a good rough measurement of the age of the universe, but cosmological observations from the last few years now allow this estimate to be greatly refined. The next and final chapter describes some of these wonderful new data sets and how they can be used to estimate precisely when the Big Bang actually happened.

SECTION 11.7: FURTHER READING

For a general introduction to cosmology and the expansion of the universe, see Roger A. Freedman and William J. Kaufmann *Universe,* 6th ed. (Freeman and Co., 2001). However, new cosmological observations are being made very quickly these days, so no book is completely up to date. Review articles in various journals and magazines are the best way to keep up. Some reasonably accessible recent articles (with references) are W. L. Freedman and M. S. Turner "Cosmology in the New Millennium" *Sky and Telescope*, October 2003, and "Four Keys to Cosmology" by various authors in *Scientific American*, February 2004. For more technical articles, www.arxiv.org is a good place to find the most recent experimental results and theoretical speculations. Also, there are some reasonably recent texbooks on cosmological subjects, including A. Liddle *An Introduction to Modern Cosmology* (John Wiley, 2003), and S. Dodelson *Modern Cosmology* (Academic Press, 2003).

The intrepid reader interested in learning how to solve problems in general relativity might want to try S. Carroll *Spacetime and Geometry: An Introduction to General Relativity* (Addison-Wesley, 2003), J. Hartle *Gravity: An Introduction to General Relativity* (Addison-Wesley, 2003), or B. Schutz *A First Course in General Relativity* (Cambridge University Press, 1994).

For measuring distances with cepheids, see W. Freedman et al. "Final Results from the *Hubble Space Telescope* Key Project to Measure the Hubble Constant" *Astrophysical Journal* 553 (2001): 47–72.

For the latest on Type Ia supernova measurements, see the following (all of which can also be found on arxiv): R. Knop et al. "New Constraints on Ω_M, Ω_Λ, and w from an Independent Set of 11 High-Redshift Supernovae Observed with the Hubble Space Telescope" *Astrophysical Journal* 598 (2003): 102–137; A. Riess et al. "Type Ia Supernova Discoveries at $z > 1$ from the Hubble Space Telescope: Evidence for Past Deceleration and Constraints on Dark Energy Evolution" in *Astrophysical Journal* 607 (2004): 665–687; John L. Tonry et al. "Cosmological Results from High-z Supernovae" *Astrophysical Journal* 594 (2003): 1–24; P. Astier et al. "The Supernova Legacy Survey: Measurement of Ω_M, Ω_Λ, and w from the First Year Data Set" *Astronomy and Astrophysics* 447 (2006): 31–48; and A. Riess et al. "New Hubble Space Telescope Discoveries of Type Ia Supernovae at $z > 1$" (2006), on-line at www.arxiv.org/abs/astro-ph/0611572.

Also useful are the websites of the supernova search teams, including the Supernova Cosmology Project at http://panisse.lbl.gov and the High-Z Supernova Search at http://cfa-www.harvard.edu/cfa/oir/Research/supernova/HighZ.html.

CHAPTER TWELVE

Parameterizing the Age of the Universe

Like Egyptologists' efforts to precisely determine the age of the pyramids, cosmologists' quest to measure the age of the universe goes far beyond a desire to have another number to stick in textbooks. It is instead part of a much larger effort to understand the large-scale structure and composition of the universe. If we take general relativity seriously—which we must, given the available data—then there is a direct relationship between the material content of the universe and the geometry of time and space. The expansion history of the universe therefore both depends upon and provides information about the properties of the materials that exist within it. There is still a lot of mystery and debate surrounding the stuff that fills our universe. Astronomical observations indicate that most of the material in the universe is *not* composed of familiar atoms, nuclei, or electrons, but laboratory studies have not yet provided much information about any of the more exotic substances that could occupy the cosmos. By studying the history of the universe in detail with a variety of data sets, cosmologists can gain valuable insights into the makeup and perhaps even the origins of the universe.

SECTION 12.1: A CLOSER LOOK AT THE EXPANDING UNIVERSE

The supernova observations described in the previous chapter are one of the more important sources of information about the history of the universe (see Figure 12.1). These data clearly indicate that the scale factor of the universe (a measure of the average distance between galaxies) has been increasing at a more or less steady rate for billions of years. Although the available data

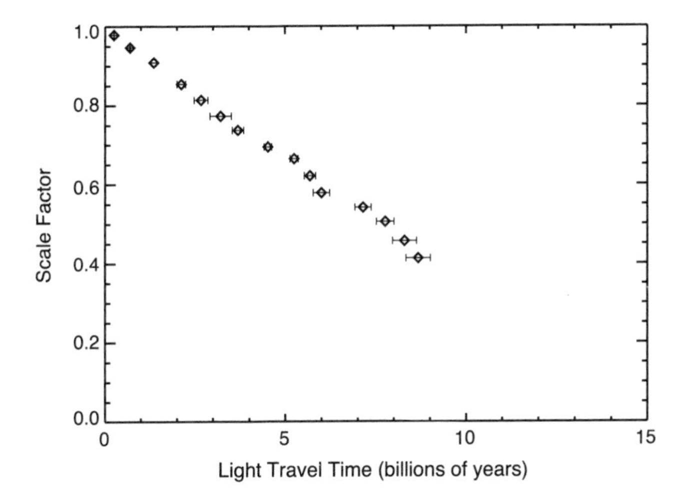

FIGURE 12.1 The scale factor of the universe versus time. This graph shows estimates of the scale factor at various times in the past deduced from supernova data from Adam G. Riess et al. "Type Ia Supernova discoveries at z > 1" *Astrophysical Journal* 607 (2004) 665–687 (available at http://www.arxiv.org/abs/astro-ph/0402512), and Adam G. Riess et al. "New Hubble Space Telescope Discoveries of Type Ia Supernovae at $z > 1$" (2006), on-line at www.arxiv.org/abs/astro-ph/0611572. This plot contains essentially the same data as those from the previous chapter, but here data from multiple galaxies have been averaged for clarity. Note that the universe is assumed to have a Euclidean spatial geometry in this plot.

go back only to a time when the scale factor was about one-half of its present value—in other words, when the average distance between galaxies was one-half of what it is today—these measurements are still able to constrain the composition and age of the universe.

Assuming that the universe is filled with a nearly uniform distribution of material,[1] the equations of general relativity yield a fairly straightforward relationship between the average amount of energy contained in each unit of volume (the energy density of the universe) and how fast the scale factor changes with time (the expansion rate). Basically, the higher the energy density of the universe, the faster the expansion rate, and vice versa.[2] The expansion history

1. Of course, material is not distributed completely evenly throughout the universe. There are objects like galaxies and clusters of galaxies that are present at some places but not in others. On suitably large scales, however, the average distribution of material is reasonably close to uniform.

2. More precisely, if the universe has Euclidean geometry, then the time rate of change of the scale factor squared divided by the scale factor squared is proportional to the average energy density of the universe.

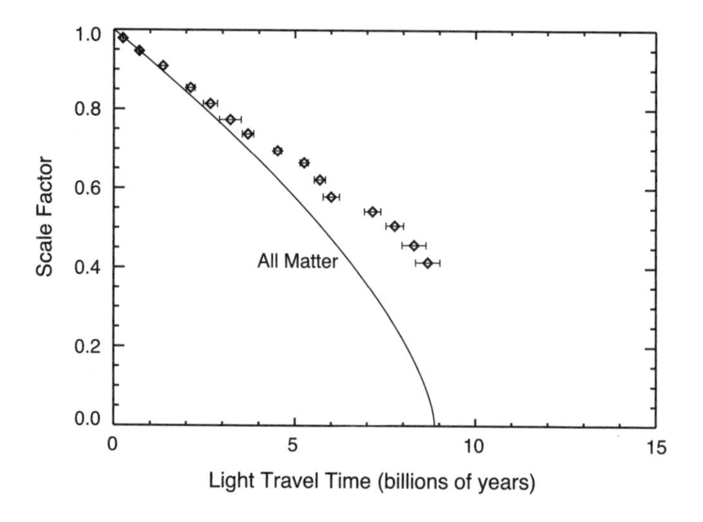

FIGURE 12.2 Observed expansion history of the universe compared with predictions assuming the universe contained only matter (assuming euclidean spatial geometry).

illustrated in Figure 12.1 therefore contains clues about the history of the universe's mean energy density. These data have revealed that the energy content of the universe has been changing in some very bizarre ways, ways that are, in fact, inconsistent with the known properties of ordinary matter.

According to Einstein's equation $E = mc^2$, most of the energy in ordinary atoms is contained in the mass of the various particles. Consequently, the energy density in ordinary matter is proportional to the number of atoms per unit of volume. As the universe expands and the scale factor increases, the average distance between particles grows, so this energy density must steadily decline with time. In fact, when the scale factor doubles, the average distance between particles increases by a factor of two in all directions, and the energy density in matter drops by a factor of eight! The energy density in matter therefore declines very rapidly as the scale factor increases. If the universe contained only matter, then this steep reduction in the energy density would strongly attenuate the expansion rate. The universe's expansion would then slow down as the scale factor increases.

Cosmologists can use the equations provided by general relativity to calculate exactly how the scale factor should have changed with time if the universe contained *only* ordinary matter like atoms. Figure 12.2 shows this theoretical prediction as a line alongside the observations from Figure 12.1. Note that this theoretical curve is steeper when the scale factor is smaller, indicating that the expansion rate was faster at that time. This illustrates how the expansion rate would slow down as the universe expands if the universe contained only atoms. The observed data diverge from this predicted trend, implying that there is something in our universe besides ordinary matter.

For astronomers, this was a bit of a shock, but it was not a total surprise. After all, they had already found evidence that there was some substance in galaxies and clusters of galaxies besides ordinary atoms. For decades, scientists have watched these distant systems, where stars and groups of stars orbit around a central point just as the planets in our solar system orbit around the sun. By measuring how fast the stars are moving, astronomers could estimate how much mass is in the galactic system. These calculations routinely show that the mass of a typical galaxy is about five to ten times as large as the total mass of all the visible stars, dust, and gas. Some astronomers believe that this is due to some feature of gravitational interactions on large scales that isn't part of standard theories of gravity. Most, however, think that these galactic systems must also contain something else in addition to the ordinary material we can see. This material does not emit light, so it has been dubbed "dark matter," but no one has yet determined exactly what this stuff is. One possibility is that dark matter is composed of exotic subatomic particles that do not interact with light or atoms in the same way that protons, neutrons, and electrons do, rendering them almost invisible. Several laboratories are currently attempting to detect such particles directly. In the meantime, astronomical observations continue to find supporting evidence for the existence of dark matter throughout the universe.

However, as mysterious as dark matter is, the supernova data call out for something even stranger. This is because—whatever it is—dark matter must become diluted as the universe expands just like ordinary atoms. If it is composed of massive particles, then the energy density is proportional to the number of particles, and as the universe expands, the energy density in dark matter will drop just as quickly as the energy density in conventional atoms. Even if dark matter is something more exotic, it must be concentrated in galaxies, or else it could not explain the observed dynamics of their gas and stars. As these galaxies are pulled apart by universal expansion, the dark matter energy density must be diluted just like the energy density in stars and gas. The theoretical prediction for a universe filled with dark matter is therefore essentially the same as the above prediction for a universe filled with ordinary atoms, and it does not match the observed data.[3]

3. In fact, cosmologists sometimes lump both dark matter and atoms together under the general name "matter."

The supernova data suggest that there must be some other form of energy, one that does not get diluted so quickly as the universe expands. In cosmological parlance, this energy must have a different "equation of state" than either ordinary matter or dark matter. People have speculated about such strange forms of energy since the very early days of general relativity. In fact, Einstein himself used a term in his equations denoted with the Greek letter Λ (lambda) that could be interpreted as one of these exotic sorts of energy. He used this parameter in his calculations because the relationship between the redshifts and the distances of galaxies had not yet been established, so he was unaware that the universe was expanding. By incorporating Λ into his equations, Einstein could describe a universe that was static—neither expanding nor contracting—which seemed more plausible to him at the time. Once he learned of the evidence that the universe was expanding, Einstein realized that he was incorporating the Λ term into his equations because of his own preconceived notions, and so called it his "biggest blunder."

Yet even Einstein's blunders have their uses, because while one form of Λ could freeze the expansion of the universe, other values and forms of this parameter have more complicated and interesting effects on the expansion rate. Such terms are now often interpreted as energy fields that have various equations of state. Over the years, however, scientists have had mixed feelings about these hypothetical fields. At some times, they seemed to provide insights into confusing and seemingly contradictory observations; and at others they appeared only to be fudge factors obscuring a deeper problem in the theory or the observations. The recent supernova data provide some of the strongest evidence yet for these oddball forms of energy, and so people have been making up fancy names for them, such as "quintessence." Unfortunately, the most popular name—and consequently the one I feel obligated to use here—for this stuff is currently "dark energy." This moniker has the potential to confuse things a great deal because it sounds far too much like "dark matter," even though dark matter and dark energy are quite different substances. Dark *matter*, as we mentioned already, is concentrated in galaxies and becomes diluted by universal expansion like ordinary matter. Dark *energy*, on the other hand, does not behave this way.

One elementary form of dark energy is vacuum energy—also known as a cosmological constant—an energy associated with space itself. If there is vacuum energy, then there is an energy per unit volume of space even if there is nothing in it. Since the amount of vacuum energy stored in any volume is unrelated to the number of particles in that volume, the density of this energy is independent of the scale factor. Therefore, if most of the energy in the universe

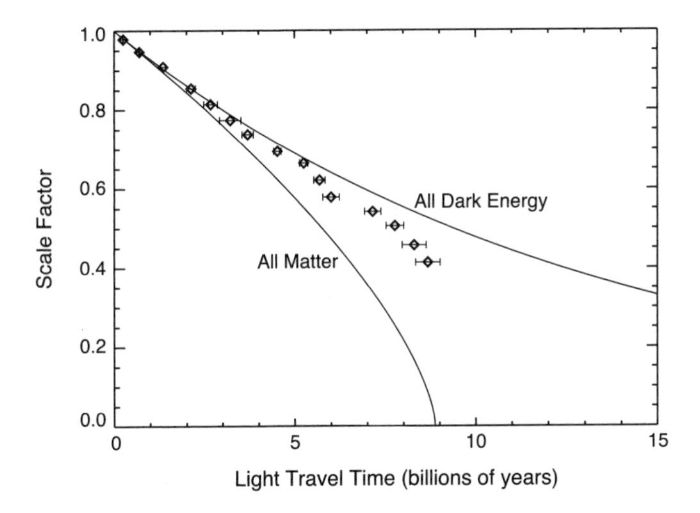

FIGURE 12.3 Observed expansion history of the universe compared with predictions assuming the universe contained only matter or only dark energy (assuming euclidean spatial geometry).

is in the form of vacuum energy, the expansion rate will not decrease as the scale factor increases. In fact, since the energy density in vacuum energy is a constant in time, the ratio of the expansion rate to the scale factor would be constant and the expansion rate would get faster as the scale factor increased.[4]

Cosmologists have postulated other forms of dark energy besides vacuum energy, and with these the energy density does not remain exactly constant with changes in scale factor. However, for simplicity's sake we will assume here that the dark energy behaves like vacuum energy, which is still a viable option. If all the energy in the universe were in the form of dark energy, then we could compute how the scale factor should evolve with time, and add this curve to the graph (Figure 12.3). In this case the slope of the curve becomes progressively steeper as the scale factor increases, which means that the expansion accelerates with time as the universe expands. While the data points fall

4. At this point you may be wondering, if energy is a conserved quantity, where is the additional dark energy coming from as the universe expands? This is a reasonable question, but the answer is far from simple. In general relativity, time and space are dynamic quantities, which complicates any effort to establish whether a given quantity is truly conserved or not. In fact, a rather big mess of math—which I must admit I have not worked through fully myself—is required to describe how energy can move through the universe. I have not yet come across a convincing intuitive way of presenting the results of these calculations, so for those readers who are enthusiastic about this subject, all I can do is point you to the general relativity course books listed at the end of this chapter; for the rest, I must beg your forgiveness.

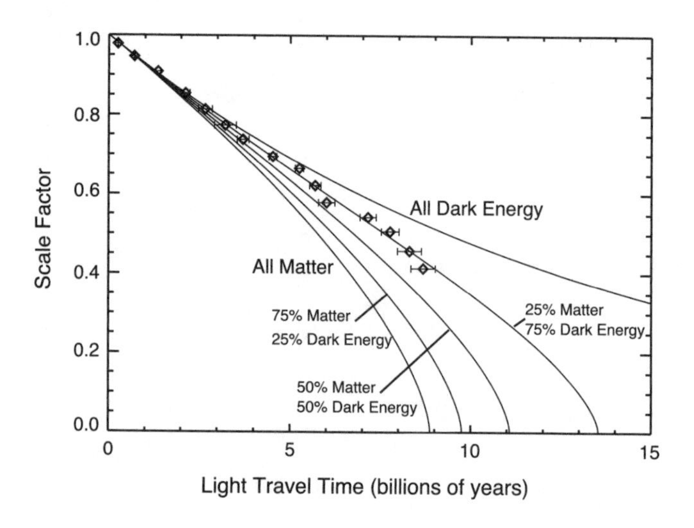

FIGURE 12.4 Observed expansion history of the universe compared with predictions assuming the universe contained various mixtures of matter and dark energy (assuming euclidean spatial geometry). The labels refer to the current proportions of matter and dark energy.

closer to this curve than to the "all matter" formulation, the fit is still not good. Notice that the data fall between these two curves, suggesting that the universe contains a mixture of dark energy and matter. Obviously, since we can observe galaxies and other celestial bodies out there, it seems clear that at least *some* matter is mixed in with the dark energy. Cosmologists can, of course, also derive predictions for a universe filled with combinations of materials, although the math is somewhat more involved because the energy density of matter declines as the universe expands, while the dark energy density stays constant. The proportions of the total energy density in these two forms will therefore change over time. While this complicates the calculations a bit, it is still possible to create well-defined theoretical curves for these different mixes of matter and dark energy, as shown in Figure 12.4 (with the labels referring to the *current* proportions of matter and dark energy in the universe). Since the data most closely follow the curve with 75% of the energy density today in the form of dark energy, we can say that our universe seems to contain about three parts dark energy for every one part matter. As we saw before, these data alone cannot tell us how much of the matter is dark matter and how much is ordinary atoms, but other observations suggest that the ratio of dark matter to ordinary matter is about five to one. Astronomers are still uncertain why the universe contains this particular mix of materials, and one might reasonably wonder whether the supernova data are somehow contaminated, thus providing a misleading picture of the expansion history of the universe. Cosmologists are well aware of this possibility, so they are always trying to use independent observations to check and refine this recipe for the universe.

In addition to simply confirming the above findings, different cosmological observations can also yield additional information about the history and the content of the universe, information that cannot be obtained readily from just the supernova data. For instance, the supernova observations alone do not provide a strong constraint on the total amount of energy contained in a unit volume of the universe. In all of the above plots, today's total energy density is assumed to have a certain value called the critical density (see below). If we decide instead that the total energy density of the universe today is really 50% higher than this, the plot would look like Figure 12.5. The theoretical curves are different from those in the previous plot because changing the energy density changes the expansion rate. But note that the data points have also shifted a bit to the right in this plot. This happened because the total energy density of the universe affects the geometry of space, and changing this parameter can alter the apparent distances to these galaxies. The calcula-

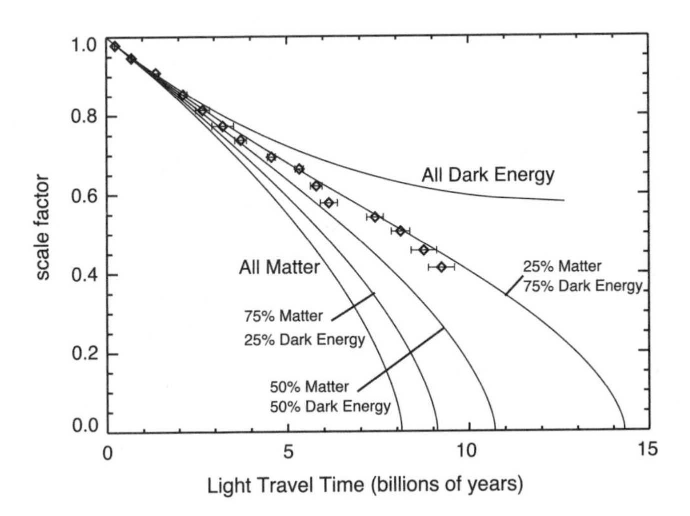

FIGURE 12.5 Observed expansion history of the universe compared with predictions assuming the universe contained various mixtures of matter and dark energy (assuming non-euclidean spatial geometry consistent with an energy density 50 percent higher than the critical density).

tions we did at the end of the last chapter to convert the apparent distances to the galaxies into light travel times are therefore sensitive to the total energy content of the universe.

In this case the data still favor the model with a 75 : 25 dark-energy-to-matter ratio. This measurement of the proportions of matter and dark energy in the universe is therefore (almost) independent of the total amount of material in the universe. However, since there is a curve that fits the data well in both plots, these observations do not tell us much about the total energy density of the universe. Furthermore, they cannot tell us precisely how long ago the Big Bang happened. Remember that the Big Bang corresponds to a time when the scale factor was zero. In the earlier plot—which assumes the energy density of the universe equals critical density—this would have happened less than fourteen billion years ago. In the later plot—representing a universe with a higher energy density—this event would have happened more than fourteen billion years ago.

A reliable account of the composition, the history, and the age of universe clearly requires more information than the supernova measurements alone can provide. Fortunately, new observations of the cosmic microwave background (or CMB) have recently yielded precise measurements of both the total energy density of the cosmos and the timing of the Big Bang.

SECTION 12.2: THE COSMIC MICROWAVE BACKGROUND

The most important characteristics of the cosmic microwave background are encapsulated in its name. First of all, the word *microwave* specifies that the CMB is a form of electromagnetic radiation. In other words, it is in the same category as X-rays, radio waves, and visible, ultraviolet, and infrared light. All of these phenomena can be treated as electromagnetic waves with different wavelengths (see Figure 12.6). X-rays have the shortest wavelengths (around 1 billionth of a meter), and radio waves have the longest (meters to kilometers). Microwaves fall towards the long end of this range, with wavelengths ranging from roughly a millimeter to 10 centimeters. This is longer than infrared light, but shorter than the radio waves that carry typical television and radio broadcasts.

Telescopes that can detect microwave radiation observe a signal from outer space from every single point on the sky.[5] This *background* of microwave radiation is (almost) constant across the entire sky, and appears to fill all of space.

5. This signal also appears in household TV sets connected to an aerial antenna, as a small fraction of the snow you see on the channels between active stations.

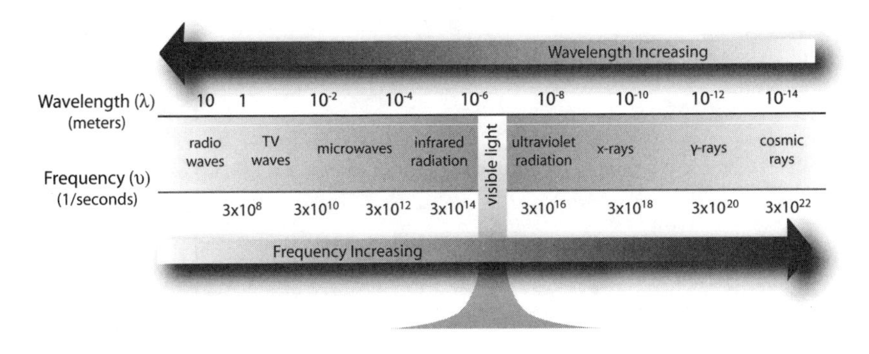

FIGURE 12.6 The electromagnetic spectrum of radiation.

Its distinctive spectrum (Figure 12.7) has a broad peak, which is reminiscent of the thermal emission from stars. Indeed, the shape of this spectrum corresponds almost perfectly with the theoretical spectrum of thermal radiation from a blackbody, an object that absorbs all the light shining upon it. The spectrum of light emitted by such an object can be computed exactly without knowing anything specific about its composition. All we need to know is the temperature of the blackbody: the lower the temperature, the longer the wavelength of the peak in the spectrum. For this microwave background spectrum, the peak occurs at wavelengths around one millimeter, which is thousands of times longer than the typical wavelengths of the peaks in stellar spectra (see chapter 10). The apparent temperature of the CMB is therefore far, far lower than the surface temperatures of stars, and in fact it is only a few degrees above absolute zero.

It is quite unusual for such a perfect blackbody spectrum to have such a low effective temperature. By definition, in order for an object to produce blackbody radiation, it must have been produced by material that interacts strongly with a broad range of electromagnetic radiation. Such materials do exist in nature. For example, plasmas, which are made up of free subatomic particles like electrons and protons with net electric charges, couple strongly to electromagnetic waves and generate thermal radiation that closely approximates a blackbody spectrum. However, plasmas exist only at very high temperatures. At low temperatures free charged particles are usually found bound into atoms, which have no net electric charge and absorb or emit light only at certain special wavelengths, producing lines like those often seen in galactic spectra (see the previous chapter). This means that blackbody radiation can usually be formed only in material that is very hot. Metals and other

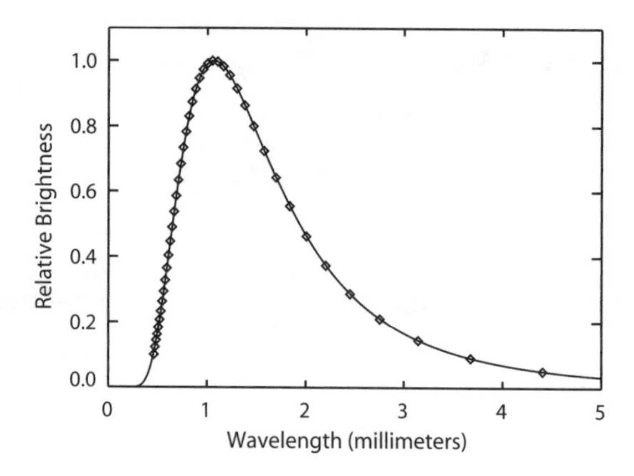

FIGURE 12.7 The spectrum of the cosmic microwave background. The points show the observed brightness of the CMB at different wavelengths. The curve is the theoretical spectrum of the thermal radiation from a 2.7 Kelvin blackbody. The differences between the observed data and the theoretical curve are almost impossible to see on this scale. (Data from D. Fixsen et al. "The Cosmic Microwave Background Spectrum from the Full COBE FIRAS Data Set" *Astrophysical Journal* 473 (1996): 573–586.)

solids can produce blackbody-like radiation at lower temperatures, but the dust and hydrogen gas in the universe almost certainly would not generate such a perfect blackbody spectrum at just a few degrees above absolute zero.

The key to resolving this paradox is to recall that we live in an expanding universe. As the universe expands, the distance between the photons increases, so the radiation is spread over an increasingly large volume. At the same time, the wavelength of every single photon gets longer (again, see the previous chapter). Both of these effects cause the spectrum of the CMB to change as the universe expands. Working through the math, we find that radiation with a blackbody spectrum at one time preserves its characteristic shape throughout the expansion, but its apparent temperature steadily decreases. If the scale factor of the universe doubles, the apparent temperature of the CMB is cut in half. As a result, the temperature of the CMB should have been higher in the past, when the scale factor was smaller.[6] In the very distant past, the effective temperature of the CMB would even be consistent with the temperature of a plasma.

6. Such changes in the temperature of the CMB can actually be observed by studying the microwave radiation in the vicinity of distant clusters of galaxies.

The *cosmic* microwave background therefore appears to be a relic from the hot, dense phase of the early universe. Shortly after the Big Bang, the universe was extremely hot and filled with a plasma consisting of free electrons and nuclei, as well as high-energy radiation like X-rays and ultraviolet light. If an electron and a nucleus combined to form an atom of neutral hydrogen, a high-energy photon would quickly come along and break the atom back into its component parts. As the universe expanded, it cooled: these disruptive photons became more spread out and their wavelengths became longer and longer. About 400,000 years after the Big Bang, there was simply not enough ultraviolet radiation remaining to keep the universe in an ionized state. At this time, electrons and nuclei were able to combine into neutral atoms and the universe came to be filled with transparent hydrogen gas. Cosmologists refer to this point in the history of the universe as decoupling, since at this time the photons stopped interacting strongly with the matter in the universe. Instead, these bits of light began to travel in roughly "straight" lines (where the definition of "straight," of course, depends on the large-scale geometry of the universe). By now, this radiation has been traveling in this manner for billions of years. In the meantime, the photons have redshifted by a factor of 1,000, moving past visible and infrared wavelengths all the way into the microwave range.

Since these CMB photons have traveled in roughly straight lines from the era of decoupling until today, the radiation that comes to us from different directions in the sky comes from different regions of the early universe. Any variations in the characteristics of the CMB from point to point across the sky therefore correspond to variations in the structure of the early universe. This means that the CMB provides us with a picture of what the universe was like when it was less than one percent of its current age.

A small but dedicated group of astronomers has been searching for these elusive variations (or anisotropies) ever since the CMB was first discovered over forty years ago. However, it was only after the launch of the Cosmic Background Explorer satellite in 1992 that small variations in the brightness were clearly detected. Since then, several ground- and balloon-based experiments have improved on these measurements. More recently, in 2003 and 2006 the data from Wilkinson Microwave Anisotropy Probe (WMAP) spacecraft provided remarkably precise measurements of the brightness variations of the CMB. It has taken so long to get these measurements because the variations are extremely small—only one part in 10,000—and only specialized cryogenic instruments can detect them.

The patience and dedication required to make these measurements have been amply rewarded by the wealth of information that they have provided about the structure of the early universe. The areas where the CMB appears slightly brighter than average correspond to regions of the universe that were a tiny bit warmer than average, while the areas where the CMB is a little dimmer correspond to regions that were cooler than average. These subtle variations in temperature reflect fluctuations in the density of the early universe. Gases and plasmas heat up when they are compressed and cool down when they expand, so the brighter regions of the CMB also correspond to denser regions of the early universe. Eventually, some of the material in these regions collapsed to form galaxies, clusters, and other objects like those we see in the universe today. These brightness fluctuations therefore provide a glimpse of the large-scale structure of the universe when it was in its infancy.

In addition to providing information about the origins of structure in our universe, the fluctuations in the CMB can also provide powerful insights into the overall composition, geometry, and age of the universe. The CMB is such a rich vein of cosmological information because the dynamics of the early universe are much simpler than those of galaxies or clusters of galaxies. These newer objects possess a broad range of densities, from nearly empty space to crowded agglomerations of stars and gas and dark matter. When this is the case, the densest regions have a disproportionate affect on their surroundings, producing complex astrophysical structures like knots, filaments, and sheets. The physics involved in the formation of such systems is so complex that large computer simulations are often needed to untangle what is going on. By contrast, the variations in the temperature and density of the early universe were extremely small. This means that once scientists have accomplished the difficult task of detecting them, the relevant equations needed to describe the variations can be well approximated with very simple expressions. Cosmologists can therefore compare the observations with theoretical predictions in a relatively straightforward way without relying too much on complex computer simulations. Furthermore, because the variations are so small, the conditions in any one region will not exert a disproportionate affect on the regions around it, so complex structures do not arise. This is confirmed by the WMAP image of the microwave sky shown in Figure 12.8, which shows no strong evidence of arcs, streaks, or any other such coordinated patterns. Instead, we see nothing more than a random distribution of bright and dark blobs of various sizes.

If we take a closer look at this image, we can see that there seems to be a characteristic scale to these bright and dark spots. In particular, the most prominent and common blotches appear to be around about one half of a degree across. If the CMB fluctuations were visible to the naked eye, such

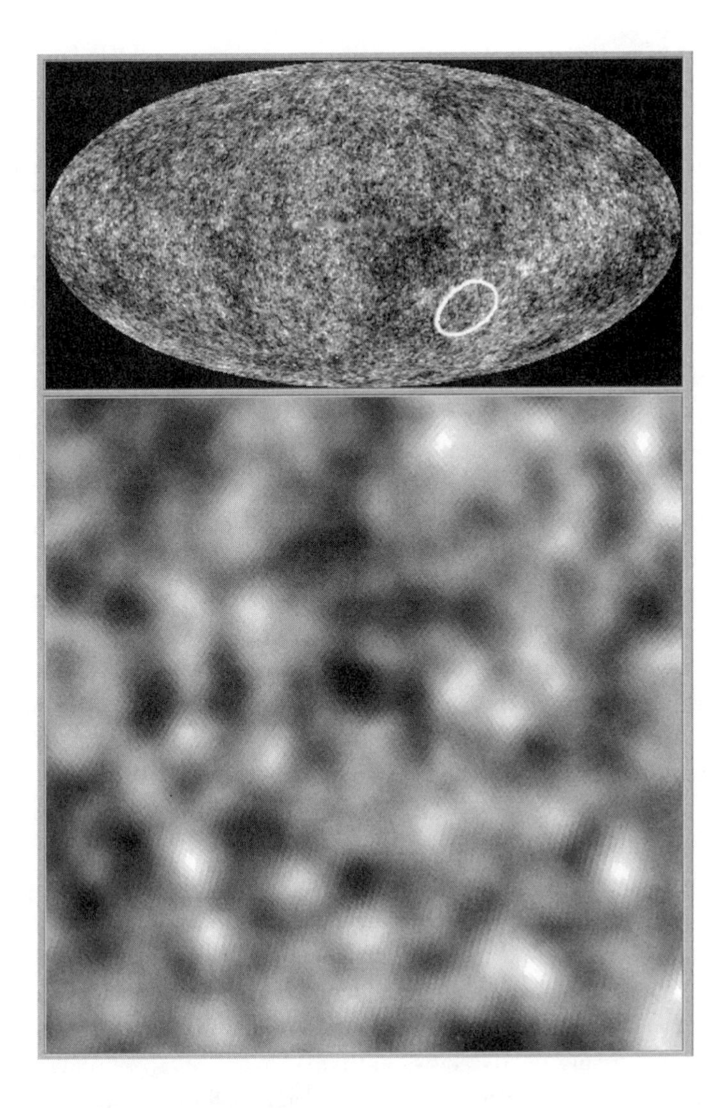

FIGURE 12.8 Variations in the brightness of the CMB measured by the WMAP satellite (this is the processed data from Max Tegmark, available at http://space.mit.edu/home/tegmark/wmap.html, displayed using the mapview software available at http://lambda.gsfc.nasa.gov/product/map/current/m_sw.cfm). The various shades of gray indicate variations in the brightness, with the brightest regions being 400 microkelvins (about one part in 10,000) brighter than the darkest regions. The top image depicts the entire sky (the Milky Way would appear as a horizontal band across the middle of this image), while the bottom image shows a close-up of the circled region (~10° across). Note that the bright and dark spots have a characteristic size, which is about half a degree on the sky.

spots would appear to be about the same size as the sun or the full moon. Of course, just because the sun, the moon, and the splotches in the CMB appear to be the same size to us here on earth, it does not necessary follow that they are all actually the same distance across. The sun is 400 times as wide as the moon, and it appears to be the same size to us only because it is 400 times as far away. By the same token, since the light from the sun takes just a little over eight minutes to reach us, while the photons that make up the CMB have been traveling through space for billions of years, the fluctuations in the CMB must correspond to very distant and very large structures in the early universe. If we knew the age and the composition of the universe precisely, then we could calculate the actual size of these features based on how far the CMB photons have traveled and how much the universe has expanded while the photons were in transit. Conversely, if we knew the actual size of the structures in the CMB, then we could infer something about the age and composition of the universe.

Amazingly, cosmologists are actually able to estimate how large these warm and cool regions were. They can do this because the patterns in the CMB anisotropies contain clues about the processes responsible for shaping the density variations in the early universe. These patterns are not always so obvious in the images of the CMB, but they become clear when the data are processed to produce a so-called power spectrum (see Figure 12.9). Unlike a normal spectrum, which shows the amount of light an object emits at different wavelengths, this power spectrum indicates how much the brightness varies on different angular scales on the sky. The tallest peak of this curve occurs at angular scales of about one degree, which simply reflects the fact that the most prominent and common bright and dark spots are about half a degree across.[7] However, the curve also contains several other peaks at progressively smaller angular scales. These reflect an excess of bright and dark spots in the CMB that are a quarter of a degree across, an sixth of a degree across, and so on.

There is currently only one plausible explanation for the regular pattern of bumps observed in the CMB power spectrum: acoustic oscillations in the plasma that filled the early universe. Imagine we had a slightly overdense (warm) region and a slightly underdense (cool) region in the primordial plasma. Material will naturally tend to move from the more dense region to the less dense region. At first, this reduces the difference in the densities. However, once the

7. The factor of two difference between the location of the peak in the power spectrum and the typical size of splotches arises because a full cycle in brightness covers both one bright spot and one dark spot.

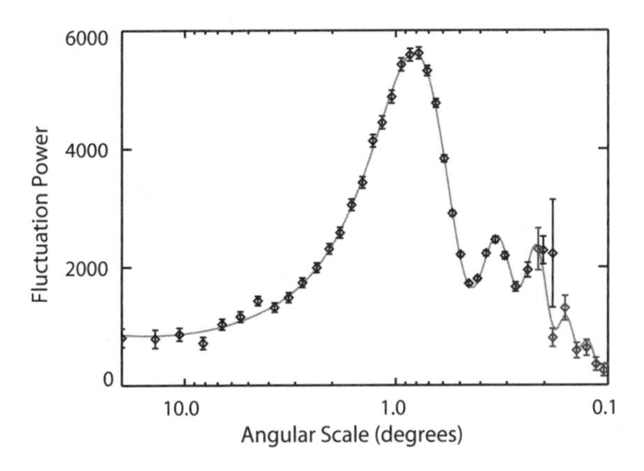

FIGURE 12.9 Power spectrul of brightness variations in the CMB. The magnitude of the brightness variations at different angular scales measured by WMAP (black points) and a ground based experiment (ACBAR, gray points). Note that smaller angular scales are at the right. As we go to smaller and smaller angular scales, the magnitude of the brightness variations rises and falls regularly; this is a signature of the ebb and flow of plasma in the early universe. Image derived from data provided by the WMAP Science Team, available at http://lambda.gsfc.nasa.gov/product/map/current; the vertical scale is in units of microkelvins squared.

material begins to flow, inertia comes into play and the previously underdense region winds up with more plasma than the formerly overdense region. Material then begins to flow back to where it came from, inaugurating a cycle in which the plasma sloshes back and forth. This oscillation causes the variations in the temperature of the plasma to repeatedly grow and shrink with time.

The time it takes for the material to flow back and forth depends on the distance between the overdense and underdense regions. The longer the distance, the longer it takes to complete an oscillation. This means that variations on different scales will be at different points in the above cycle during the era of decoupling. Say hypothetically that the plasma has just enough time prior to decoupling to flow out from an overdense region and accumulate in a previously underdense region 400 thousand light-years away. This means there will be a large density variation between those points, and a large temperature difference will be imprinted on the CMB. Similarly, if the two points were only 200 thousand light-years apart, then the material would have had time to flow out of the overdense regions and then come back. Again, there would be a significant density variation between the regions at decoupling. By contrast, if the two regions were 300 thousand light-years apart, the plasma would still be flowing between the two regions, and the density variations at decoupling would be more subdued.

Now imagine there was an array of overdense and underdense regions in the early universe with a broad range of separations. At decoupling, only the regions separated by a specific, regularly spaced set of distances (such as 400,000 light-years, 200,000 light-years, 133,333 light-years, etc.) will display large density variations. The variations in the CMB will then be enhanced at a similarly regular set of angular scales. This is exactly the pattern we see in Figure 12.9, so we have very good reasons to believe that this ebb and flow of plasma in the early universe really happened.

In this scenario, the various peaks in the power spectrum of the CMB correspond to distances in the early universe where the plasma could complete 0.5, 1, 1.5, or 2 cycles of its journey back and forth between overdense and underdense regions. These distances are determined by two factors: how fast density variations in the plasma can move and how much time is available for the plasma to oscillate before decoupling. For high-temperature plasmas, which are rich in photons, variations in the density of the plasma such as these should propagate at speeds comparable to the speed of light.[8] Since the speed of light is a well-measured constant of nature, we are able to calculate with a high degree of accuracy how fast these density perturbations could change. This means that if we can determine how much time elapsed between the Big Bang and decoupling, we can determine the distances that correspond to the peaks in the CMB power spectrum.

Strangely enough, it is easier to estimate the age of the universe at decoupling than it is to estimate the age of the universe today. The latter is a tricky business because we need to have a fairly comprehensive knowledge of how much matter and dark energy there is in the universe. However, during the time before decoupling, the exact concentrations of dark matter and dark energy should be much less relevant. Remember that the energy density in dark energy does not change as the universe expands, while the energy density in matter decreases. In the distant past, when the scale factor was much smaller and particles were packed much more closely together, the energy density in matter would have been much higher than it is today, and the dark energy would then be a much smaller fraction of the total energy in the universe. The dark energy therefore probably had a negligible effect on the expansion rate at this time. Similarly, the energy density in light gets diluted even faster

8. More precisely, they move at about 60% of the speed of light.

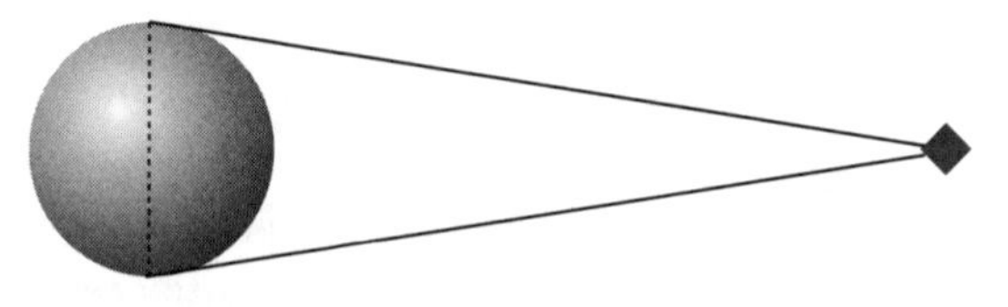

FIGURE 12.10

than the energy density in matter—the wavelength of photons increases at the same time the photon density declines—so at sufficiently early times the energy density in radiation was much higher than everything else. It turns out that for most of the time prior to decoupling, the energy density in radiation was dominant, and since the intensity of the CMB tells us precisely how many photons there are in the universe today, there is not much uncertainty in the age of the universe at decoupling.[9] Working through the math, cosmologists find that decoupling occurred about 400,000 years after the Big Bang. Combining this information with the fact that the density variations move at speeds comparable to the speed of light, we find that the half-degree wide blobs we see today in the CMB were 400,000 light-years across at decoupling.

In most situations, once we know both how big something is and how big it appears to be from a particular vantage point, we can figure out how far away it is. Imagine the two sight lines to opposite sides of an object, forming two sides of a triangle, as in Figure 12.10, where the diamond on the right represents the observer. The length of the base of this triangle is the physical size of the object, and the angle between the two long sides is determined by how big the object appears to be. With these two numbers and a few basic rules of geometry we can normally calculate the height of the triangle and the distance between the object and us. We could even figure out how long it would take light to travel from the object to us. Similarly, we can use the apparent size of the 400,000-light-year-wide blobs in the early universe to determine how long the CMB photons have been traveling through the universe. However, this calculation is not as simple as we might at first expect. Not only has the universe expanded by a large factor since decoupling, the large-scale geometry of the universe may not follow the familiar Euclidean rules.

9. The actual calculation, of course, cannot completely neglect the amount of dark and ordinary matter in the universe, or the nonnegligible contribution of neutrinos, but these complications do not greatly increase the uncertainty in the calculations.

In the last chapter, we saw how general relativity interprets gravity as a distortion in the geometry of space and time that depends on the amount of matter and energy in the area. We also saw how this interplay between matter and geometry can explain the data found in a Hubble diagram. However, the total energy density of the universe does not just alter the distances between galaxies, it also determines a fundamental aspect of the geometry of the universe known as the curvature. If the curvature of the universe is zero, the precepts of Euclidean geometry apply: parallel lines never intersect, the sum of the internal angles of a triangle is always 180 degrees, and so on. By contrast, if the curvature is anything but zero, then a different set of rules apply, rules that have more in common with the geometry of curved surfaces. If the universe has a positive curvature, then its geometry would be like that of the surface of a sphere or a globe. Here you can easily make a triangle using two lines of longitude and one line at the equator where every internal angle would be 90 degrees, and their sum would far surpass the 180 degrees required in Euclidean geometry. Conversely, if the universe has a negative curvature, it would have properties similar to hyperbolic surfaces, where the interior angles of triangles total less than 180 degrees.

The curvature of the universe is dependent upon its total energy density. If the energy density equals a special value called the critical density, the curvature of the universe will be zero. If the universe had exactly zero curvature at one time, it should never develop positive or negative curvature. The universe would then have Euclidean spatial geometry now and at any point in the past.[10] However, if the energy density is higher than the critical density, then the curvature of the universe is positive; conversely, if the energy density falls below the critical density, then the curvature is negative.

Because the spatial geometry of the universe determines the shape of the paths that light takes as it travels through the universe, the curvature affects the apparent sizes of objects seen from a distance. For example, below we have an illustration of three observers viewing the same sized object in universes with different geometries:

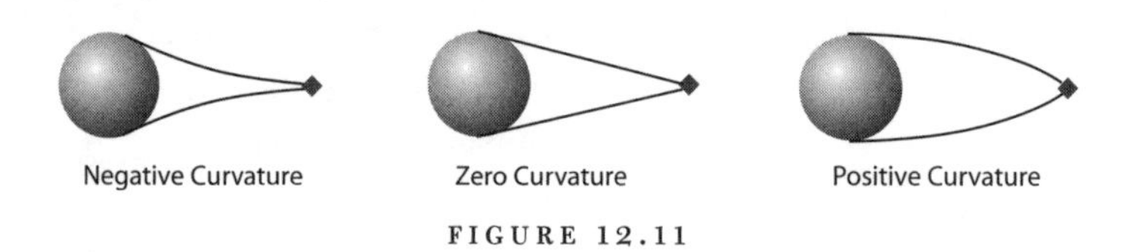

FIGURE 12.11

10. Note that since the curvature remains zero as the universe expands, the value of the critical density usually changes with time.

In the middle panel, the geometry has zero curvature, and light travels along truly straight lines to produce a familiar Euclidean triangle. In the other two cases, the curvature of the universe means that the light must follow somewhat curved paths. Suppose that the angle between the two rays at the observer's position in the zero curvature case is 10 degrees, then the object would appear to be 10 degrees across. In the negative curvature case this angle is smaller, causing the object to appear less than 10 degrees across. Finally, in the positive curvature case, the angle is larger, so the object will seem to be more than 10 degrees across. The same object therefore appears bigger when the curvature is positive and smaller when the curvature is negative.

Were we to move these objects so that all three observers saw the object as 10 degrees across, the effect of the curvature on the apparent distances becomes obvious. Since the object appeared bigger than this in the positive curvature case, it must now be carried farther away. Likewise, since the object in the negative curvature case appeared to be smaller than 10 degrees, it now must be brought closer to the observer:

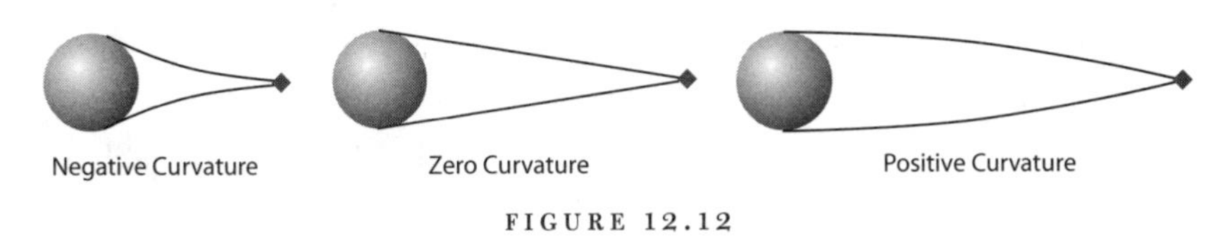

FIGURE 12.12

This shows us that if the observers assume geometry follows Euclidean rules, they can misjudge the distance to the object. Objects appear closer than they actually are if the curvature is positive, further away if the curvature is negative.

In a similar way, if bright regions 400,000 light-years across at decoupling appear to be half a degree wide today in the CMB, the distance traveled by the light between decoupling and today depends on the curvature of the universe. If the curvature is positive, then the light has a longer distance to travel; if the curvature is negative, then this distance is shorter. Therefore, if we knew the actual distance the light traveled, we could figure out the curvature of the universe and in turn infer its total energy density. Conversely, if we knew the total energy density and the curvature of the universe, we could calculate how far the light has traveled and then use this information to determine the age of the universe. While it might seem that we will need to make some additional measurements before we can deduce anything about the age or geometry of the universe, the CMB alone actually provides enough data for cosmologists to uncover both the curvature and the distance the photons have traveled. This is because the total energy density determines both the overall geometry of the universe and how the universe expands between decoupling and today.

At first it might appear that the expansion of the universe would just add another level of complication to this problem, since we now have to relate the angles measured today to a physical distance from the distant past, when the scale factor of the universe was much smaller. However, such issues are not major obstacles for this particular analysis because the spectrum of the CMB tells us exactly how much the universe has expanded since the era of decoupling. Again, the comparative simplicity of the early universe is the key, for it enables cosmologists to calculate the spectrum of the light released from the decoupling plasma reliably. Remember that at this time the universe was undergoing a phase transition: the primordial plasma of free charged particles was transforming into neutral hydrogen gas. This transition occurred at temperatures of a few thousand degrees Celsius and when the average distance between hydrogen atoms was about a millimeter. Similar conditions can be achieved in a laboratory, so the physics of this transition are well known and the spectrum of the thermal radiation released at decoupling can be computed accurately. It turns out that this light had a typical wavelength of about one micron, or a thousandth of a millimeter, which is similar to the typical wavelengths of most starlight (this isn't too surprising since both are the light released from the "surfaces" of hydrogen-rich plasmas). Today, the peak of the CMB's spectrum lies at wavelengths around a millimeter, or about a thousand microns, so the universe must have expanded by a factor of about 1,000 between decoupling and today. We can therefore be reasonably confident that regions 400,000 light-years across at decoupling would be a little over 400 million light-years across today, and consequently there is little uncertainty in how the present-day angle measurements should relate to these ancient distance scales.

In fact, far from being a nuisance, the expansion of the universe is an essential component of the calculation of the universe's geometry and age because the way the universe expands determines how far light is able to travel after it is released from the decoupling plasma. Remember that light travels at a well-defined, finite speed of about 300,000 kilometers per second. Light can therefore only travel a finite distance during the time between decoupling and today. The equations of general relativity tell us that higher energy densities correspond to faster expansion rates, so the total energy density of the universe affects how long it takes for the universe to expand by a factor of 1,000, which in turn constrains how far light can travel during this time. Of course the exact relationship between this distance and the total energy density depends somewhat on the composition of the universe, but so long as the ratio of dark energy to matter is not too large, a universe with higher total energy density will expand more quickly at early times, and this more rapid expansion will in turn reduce the distance light can travel during the time between decoupling and today. Now, a higher energy density also makes the curvature of the universe more positive, which means the decoupling plasma would have to be farther away for the 400,000-light-year-wide splotches in the CMB to appear half a degree across. Increasing the energy density therefore tends to *increase* the distance the photons *must* travel to be consistent with the observations while it *decreases* the distance light *can* travel in the time allotted. Similarly, a lower value of the energy density would increase the range of the CMB photons while decreasing the distance they must travel. Thus there is only one value of the energy density that will result in the photons traveling the required distance in the allowed amount of time. After going through all of the relevant calculations, cosmologists find that (almost regardless of what they assume about the universe's composition) the energy density of the universe must be close to the critical density, which means the curvature of the universe has to be near zero. The full implications of this discovery are still not completely understood, but many cosmologists think that this lack of curvature may be an important clue about the state of the universe during the very first moments after the Big Bang.

Such speculations aside, knowing the curvature of the universe also points us towards a precise measurement of the age of the universe. Since the curvature is zero, we may use the standard rules of Euclidean geometry and—after accounting for the actual expansion history of the universe—calculate how far the light has traveled since decoupling. Then, given the speed of light, we can also estimate how long the light has been traveling, which tells us how much time has elapsed since decoupling. Finally, we can add the age of the universe at decoupling (a trivial correction of 400,000 years) and we have the age of the universe: 13.7 billion years old, give or take a few hundred million years.[11]

Obviously, this figure relies on several assumptions about the nature and early history of the universe. Cosmologists are currently working to test these assumptions and to refine our understanding of the composition and the history of the universe. New measurements of Type Ia supernovae and the CMB are now under way in order to confirm the accuracy of the available data and

11. The small uncertainty in this number is in part a side effect of the universe's lack of curvature, which makes the age estimate rather insensitive to residual uncertainties in the composition of the universe, etc.

to reduce the uncertainties on parameters such as the curvature of the universe and the mix of matter and dark energy. Meanwhile, other cosmological data sets are providing ways to verify and supplement these findings. For example, the globular cluster data described in chapter 10 indicates that the oldest stars in the universe are around 13 billion years old. Given the relatively large uncertainties in the measurements, these ancient stars are not too old for a universe that formed 13.7 billion years ago, so these data are consistent with the results derived from the CMB. Indeed, the CMB and the supernova data are currently compatible with a broad range of cosmological observations, including measurements of the total mass in clusters of galaxies and the mix of elements and isotopes in the early universe. All of these data sets therefore seem to support the idea that the universe contains both dark matter and dark energy, has zero curvature, and is about 13.7 billion years old.

As these cosmological measurements are improved and scrutinized over the next few years, they could corroborate and refine our current understanding of how the universe operates. For example, they could reveal whether the dark energy is really an energy associated with empty space or some other type of energy field. Alternatively, these data could reveal some unforeseen phenomenon that profoundly alters our ideas about the universe. Just like all of the subjects covered in this book—and many other areas of scientific and scholarly work—what is most exciting about this field is the potential for major advances and unexpected discoveries in the not too distant future.

SECTION 12.3: FURTHER READING

As with the previous two chapters, a good book for getting an overview of astronomy and cosmology is R. A. Freedman and W. J. Kaufmann *Universe,* 6th ed. (Freeman and Company, 2001). Some nice articles (with references) about new developments is cosmology are W. L. Freedman and M. S. Turner "Cosmology in the New Millennium" *Sky and Telescope*, October 2003, and "Four Keys to Cosmology" in the February 2004 issue of *Scientific American*.

For more details on the WMAP satellite, see its website, http://map.gsfc.nasa.gov/; see also and the archive site where the data are available: http://lambda.gsfc.nasa.gov/. For more technical discussions of the nature of dark energy and dark matter, I strongly suggest perusing the on-line archive of astrophysical articles at www.arxiv.org.

Those readers interested in learning the intricacies of general relativity might want to look in the textbooks S. Carroll *Spacetime and Geometry: An*

Introduction to General Relativity (Addison-Wesley, 2003), J. Hartle *Gravity: An Introduction to General Relativity* (Addison-Wesley, 2003), B. Schutz *A First Course in General Relativity* (Cambridge University Press, 1994), and S. Dodelson *Modern Cosmology* (Academic Press, 2003).

Those readers masochistic enough to want to understand how to use general relativity to predict the power spectrum of the CMB should read H. Kodama and M. Sasaki "Cosmological Perturbation Theory" *Progress of Theoretical Physics Supplement* 78 (1981).

Those readers who want a more approachable introduction to CMB anisotropies should see the websites http://background.uchicago.edu/~whu/beginners/introduction.html and www.astro.ucla.edu/~wright/cosmolog.htm and the links therein.

GLOSSARY

Absolute Magnitude: The magnitude that a star would have if it were located 32.6 light-years away.

Absolute Zero: The lowest possible temperature, where the random thermal motion of atoms would be effectively zero.

Accelerator Mass Spectrometry (AMS): A type of mass spectrometry that uses large particle accelerators and multiple stages of deflection to cleanly isolate rare isotopes.

Achondrite: A meteorite that is composed primarily of silicate minerals and lacks chondrules.

Acoustic Oscillations: The ebb and flow of plasma between overdense and underdense regions in the early universe. Such motions are probably responsible for the observed brightness variations in the CMB.

Afrotheria: A group of placental mammals, primarily found in Africa, that includes elephants, hyraxes, manatees, aardvarks, tenrecs, and golden moles. Molecular analyses indicate these animals all belong to a common branch of the mammalian family tree.

Aluminum-26: An unstable isotope of aluminum with a half-life of 730,000 years. It is used to estimate the sequence of events that occurred in the early solar system.

Alpha Decay: A form of nuclear decay where the nucleus splits into two pieces, one of which consists of two protons and two neutrons (a helium nucleus).

Amino Acid: A type of molecule that is the basic building block of proteins. The twenty different types of amino acids each have different chemical properties that affect the shape and functionality of the protein.

Apparent Magnitude: The magnitude of a star observed from earth.

Ardipithecus ramidus: A fossil hominid from Ethiopia that may have lived over five million years ago. More complete remains of this animal could provide information about how and when our ancestors became bipedal.

Australopithecus afarensis: A fossil hominid that lived about 4.5 million years ago. This is the earliest hominid that is known from a reasonably complete skeleton (the famous "Lucy"). It had a brain size not much larger than that of a chimpanzee, but was already able to walk upright on two legs.

B'alah Chan K'awiil: A ruler of the classic Mayan city of Dos Pilas, who was heavily involved in the battles between Yuknoom Ch'een and Nuun Ujol Chaak.

Base Pair: A pair of complementary nucleotides (A-T or G-C), each attached to one strand of a DNA molecule.

Bayesian Statistics: A type of statistical analysis that takes the probability of getting a result given a theoretical prediction and converts it into the probability that the theoretical prediction is correct given the available data.

Beryllium-10: An unstable but long-lived isotope of beryllium produced by cosmic rays like carbon-14. Levels of beryllium-10 in ice-cores provide information on the history of the cosmic ray flux and state of the solar and geomagnetic fields over the last 20,000 years.

Beta Decay: A form of nuclear decay where a neutron converts into a proton, emitting an electron and a neutrino. Variants on this process include a proton capturing an electron and transforming into a neutron.

Big Bang: A singular event in the history of the universe, when the mean density of material approached infinity. Theoretical descriptions of the universe break down at this point so it is unclear what may have occurred at this time. The Big Bang is taken to mark the beginning of the universe as we know it.

Blackbody: An object that absorbs all of the radiation incident upon it.

Calcium and Aluminum Rich Inclusions (CAIs): Irregularly shaped regions in chondrites that have high quantities of refractory elements such as aluminum and calcium. Most were probably formed very early in the history of the solar system.

Calakmul: The modern name of an important Mayan center during the Classic period. It was originally known as *Chan*, or "snake." Yuknoom Ch'een was one of its most prominent rulers.

Calendar Round: The part of the Mayan calendar that is composed of the Tzolk'in and the Haab. It specifies when an event occurs within a fifty-two-year cycle.

Calibrated Carbon-14 Date: A carbon-14 date that accounts for the fluctuations in the carbon-14 levels in the atmosphere over time. The conversion between conventional and calibrated carbon-14 dates is performed using a standard calibration curve derived from tree rings and other sources.

Carbon-12: The most common, stable form of the element carbon, which contains six protons and six neutrons.

Carbon-13: A stable isotope of carbon, with seven neutrons.

Carbon-14: An unstable isotope of the element carbon, with six protons, eight neutrons, and a half-life of 5,730 years. It is produced by interactions between cosmic rays and nitrogen atoms in the upper atmosphere.

Carbon-14 Dating (Radiocarbon Dating): A method of measuring the age of organic material. First, the present amount of carbon-14 in a sample of material is measured. Then the original amount of carbon-14 in the sample is inferred based on data from tree rings and other sources. Finally, the age of the material is computed from the fraction of the carbon-14 that has decayed and the half-life of carbon-14.

Carbon Cycle: The flow of carbon atoms between the atmosphere, living organisms, the oceans, etc.

Caverna da Pedra Pintada: An archaeological site located in the tropical lowlands of Brazil. People here were apparently foraging for small game and eating nuts at the same time that people in North America were hunting mammoths with Clovis points.

Celestial Pole: The point on the sky that all the stars seem to move around as the earth rotates. At present, the north celestial pole lies very close to Polaris, the "pole star." However, the position of the celestial pole slowly moves in a circle through the stars as the earth precesses.

Cepheid: A luminous star whose brightness varies in a characteristic saw-tooth pattern with a period that ranges from a day to weeks. The period of the cepheid is correlated with its luminosity, making them useful distance indicators.

Chondrite: A meteorite that is composed primarily of silicate minerals and contains chondrules. They are probably relics from a very early stage in the formation of the solar system.

Chondrules: Small spheres of rock, about a millimeter across, found in most stony meteorites.

Classic Period: A period of Mayan history between roughly 250 and 900 CE, when the Mayans living in the tropical lowlands of Guatemala, Belize, western Mexico, and far eastern Honduras built large cities and erected many carved stone monuments.

"Clovis-First": A model for the settlement of the Americas that posits that the makers of Clovis points were among the first people to arrive in the New World. According to this hypothesis, people reached Sub-Canadian America through an ice-free corridor that opened up during the end of the last Ice Age. People followed large game like mastodons down this corridor and spread rapidly throughout North and South America. As the supplies of large game declined, people began to settle down and use local resources more intensively. This idea has been challenged by recent archaeological discoveries.

Clovis Points: Elaborate stone tools found throughout North America. They are large flaked projectile points, with a distinctive "flute" at their base that requires some skill to produce. These tools date back to around thirteen thousand years ago, so they are among the oldest evidence of human occupation in the New World.

Color: In astronomy, the difference in the magnitude of a source measured through two different filters.

Color-Magnitude Diagram (Hertzsprung-Russel Diagram): A plot of stars' absolute magnitude versus their color.

Continental Drift: The change in the position of the continents over time.

Conventional Carbon-14 Dates: Carbon-14 dates calculated assuming the carbon-14 content of the atmosphere has been a constant in time and the half-life of carbon-14 is 5,570 years. Used as a standardized measure of the carbon-14 content of a sample.

CMB: See Cosmic Microwave Background.

Convergence: In biology, a process whereby two organisms that are not closely related come to acquire similar traits. Typically, this happens because the ancestors of both animals were subjected to similar environmental pressures.

Coordinate Distance: In an expanding universe, the distance between two objects at a particular point in time, typically the present.

Cosmic Microwave Background (CMB): A nearly constant background of microwave radiation that appears to fill all of space and to be a relic from the hot, dense phase of the early universe. Provides important information about the composition, geometry, and age of the universe.

Cosmic Rays: Atomic nuclei and other subatomic particles that are constantly colliding with our atmosphere at very high speeds. These particles may be produced by a variety of astrophysical objects, but pinpointing their origin is difficult because they have been deflected by interstellar magnetic fields. Collisions between these particles and nitrogen atoms in our atmosphere produce carbon-14.

Cosmological Constant: See Dark Energy.

Critical Density: In cosmology, the density the universe must be in order for the universe to have zero curvature.

Curvature: A parameter that measures how much the geometry of the universe deviates from Euclidean rules. Positive curvature implies that the interior angles of triangles sum to greater than 180°, while negative curvature implies that they sum to less than 180°.

Dalton Minimum: A period around 1820 when the sun possessed an abnormally low number of sunspots.

Dark Energy (Quintessence, Lambda, Cosmological Constant): A form of energy that does not become diluted as the universe expands like matter does. Required to explain the Type Ia supernova data.

Dark Matter: A form of matter that does not produce light and is possibly composed of exotic subatomic particles.

Decoupling: The point in the history of the universe when the density of high-energy photons had dropped far enough that electrons and nuclei could combine to form atoms of hydrogen. At this time the matter in the universe undergoes a phase transition from a plasma to a hydrogen gas.

Dendrochronology: The study of the patterns in tree rings in order to determine when the rings were produced. The width of a tree ring reflects the growing conditions in a particular year of the tree's life. By comparing the distinct patterns of thickness in the rings from living and long-dead trees, it is possible to extract exactly what year each ring was laid down, going back millennia. These data provide a detailed record of local climatic conditions and atmospheric carbon-14 content.

Dendrogram: See Phylogenetic Tree.

Deoxyribonucleic Acid (DNA): A molecule found in most living cells, composed of two spiral chains connected by a set of nucleotide base pairs. Information encoded in the sequence of base pairs determines how the cell operates.

Doppler Shifts: Changes in the wavelength of sound or light due to the relative motion of the source and the observer.

Dynasties (Egyptian): The thirty-one stages into which ancient Egyptian history is conventionally divided. Each dynasty corresponds to the reigns of a certain group of pharaohs, ultimately based on the listing by the Egyptian priest Manetho. However, it is still not clear why certain groups of rulers were considered members of a single dynasty.

Electron: A negatively charged subatomic particle. Electrons are much less massive than protons or neutrons and are not confined to the nucleus. Instead, they form a diffuse cloud that interacts with similar electron clouds surrounding other atoms. The electrons therefore play a dominant role in most of the chemical reactions between atoms. The number and configuration of the electrons in an atom is determined by the number of protons in the nucleus.

$E = mc^2$: Einstein's famous equation that says that there is an energy associated with any mass.

East African Rift System: A suite of geological features extending down the east side of Africa from Eritrea to Mozambique. This region shows evidence that the earth's crust is or was recently being pulled apart. The resulting stresses created a series of valleys and gave rise to widespread volcanic activity. This region has also yielded many early fossil hominids.

Element: A type of atom with a given number of protons in its nucleus. All the atoms of a given element have basically the same chemical properties.

Energy: A quantity that can never be created or destroyed, but can be transformed into various forms. One form of energy is carried by objects in motion, with the speed (and mass) of the object determining how much of this type of energy it has. Changing the speed of a particle therefore requires converting some of this energy into another form or vice versa, so any form of energy can be regarded as the potential to produce motion.

Eomaia scansoria: The earliest known eutherian mammal, found in China within deposits 120 million years old.

Equation of State: In cosmology, the parameter that describes how the energy density of a material changes as the universe expands.

Euarchontoglires: A group of mammals discovered in molecular analyses that includes rodents, rabbits, primates, flying lemurs, and tree shrews.

Euclidean Geometry: A set of precepts that include such familiar concepts as "parallel lines never intersect" and "the interior angles of triangles always add up to 180 degrees."

Eutheria: The group of all mammals, living and extinct, that are more closely related to modern placental mammals than to any other modern animal.

Expansion, Universal: The apparent increase in the distance between galaxies over time due to a continuous change in the geometry of the universe.

Flanking Regions: Regions at the start and end of a gene that do not carry information about how to make a specific protein, but do indicate where that information is located in the gene and when it should be accessed. These regions often have characteristic sequences that allow biologists to identify genes even if they do not know what the protein encoded in the gene actually does.

Fusion: The assembly of multiple atomic nuclei into a single larger nucleus. The fusion of hydrogen nuclei into helium provides the power necessary to support main sequence stars.

Gamma Decay: A form of nuclear decay where the nucleus loses energy by emitting a photon.

Gene: A location on a DNA molecule that contains the instructions for making a protein.

General Relativity: The theory originally developed by Albert Einstein that posits that gravity is not a classical force but is instead a distortion in the geometry of space and time. This theory has strong support from a variety of observations, including the lensing of light by massive objects.

Globular Cluster: A spherical collection of up to a million stars packed into a region with a density hundreds of times as high as that found in our galactic neighborhood. The oldest stars in these objects are among the oldest that are known.

Great Apes: A group of animals consisting of humans, chimpanzees, gorillas, and orangutans.

Haab: A 365-day cycle that is part of the Mayan Calendar Round. It consists of eighteen "months" of twenty days, plus an extra period of five days.

Half-life: The time it takes half of the radioactive atoms in a given sample to decay.

Heliacal Rising: An event that occurs when a given star rises just before the sun. The heliacal rising of Sirius was used by the ancient Egyptians to mark the beginning of the year, when floods covered the Nile valley. Since the Egyptian calendar did not include leap days, this event would occur at different dates in different years.

Hertzsprung-Russel Diagram: See Color-Magnitude Diagram.

Hominids (Hominins): Those animals that are more closely related to living humans than any other living animal. All of the fossil hominids found to date share some of the traits that are now unique to humans, such as large brains and an upright posture. These fossils can tell us how and when our ancestors first acquired these traits.

Hubble Diagram: A graph that shows the redshifts of galaxies as a function of distance.

Ice Age: A period when the climate was much colder than it is now. The last Ice Age ended about 15,000 years ago.

Intermediate Periods: Three periods in Egyptian history when the central authority of Egypt was weak. Establishing the exact chronologies of these periods is difficult. In particular, uncertainties regarding the length of the First Intermediate Period make it difficult to estimate the age of the Old Kingdom based on historical records alone.

Introns: Stretches of DNA in a gene that are interspersed among the sequences that provide information for making a protein but do not themselves provide any information about the protein.

Iron Meteorites: Meteorites that are composed primarily of metallic alloys of iron and nickel.

Isochron Dating: A method of radiometric dating that uses the isotopic composition of multiple components in an object to infer the original amount of a radioactive isotope in the material and its age. Often used to date ancient rocks and meteorites.

Isochron Plot: A plot of the isotope ratios for an array of minerals from a single object. Used in isochron dating.

Isotope: Atoms of a given element with a given number of neutrons; atoms with the same numbers of protons but different numbers of neutrons are considered different isotopes of the same element. Different isotopes of a given element have almost identical chemical properties, but different masses. Some isotopes can also be stable while others are unstable.

Kelvin: A unit of temperature. A temperature difference of 1 Kelvin is the same as a temperature difference of 1 degree Celsius, or 1.8 degrees Fahrenheit. A temperature of 0 Kelvins (−273.15°C) is absolute zero.

Lambda: See Dark Energy.

Laurasiatheria: A group of mammals uncovered by molecular analyses that includes whales, both types of hoofed animals, carnivorans, pangolins, bats, moles, shrews, and hedgehogs.

Light-Year: The distance light can travel in a year, equivalent to 9.5 trillion kilometers.

Long Count: The part of the Mayan calendar that records the number of days that have elapsed since some (probably mythological) event that occurred in August 3114 BCE.

Luminosity: The total amount of energy emitted by a star in the form of electromagnetic radiation.

Luminosity Distance: A distance to an astronomical object that is calculated based on the luminosity of the object and its observed brightness.

Magnitude (of Brightness): A measure of the brightness of an astronomical body. Each unit of magnitude corresponds to a factor of roughly 2.5 in brightness, with a decrease in magnitude representing an increase in brightness. A magnitude 1 star is therefore 2.5 times as bright as a magnitude 2 star, which is 2.5 times as bright as a magnitude 3 star.

Main Sequence: The diagonal line in a color-magnitude diagram that runs from bright blue stars to faint red stars; most nearby stars fall along this line. The stars that fall along the main sequence are powered mainly by the fusion of hydrogen into helium.

Main Sequence Turn-off: The location of the bright blue end of an incomplete observed main sequence for a cluster of stars. The stars at this location are just about to convert into red giants, so the position of this feature depends on the age of the cluster.

Mammals: A group of animals that have hair, nourish their young with milk, and can regulate their body temperature. Divided into about twenty orders.

Marsupial: A mammal that keeps its young in a pouch.

Mass: A quantity associated with objects that determines how they move in response to outside forces. Einstein's equation $E = mc^2$ says that there is an energy associated with this quantity.

Mass Fractionation: A phenomenon whereby isotopes with different masses become separated from one another because some process can more efficiently transport less massive atoms than more massive atoms, or vice versa.

Mass-Luminosity Relation: The strong correlation between the luminosities of main sequence stars and their masses.

Mass Spectrometry: A method of sorting and identifying the components of a sample by mass. It uses electrostatic acceleration and magnetic deflection to separate ions of different masses, which can then be counted.

Matter: In cosmology, a generic term for any material where most of the energy is stored in the mass of particles.

Maunder Minimum: A period between 1650 and 1700 with an unusually low number of sunspots.

Meadowcroft: An archaeological site in southeastern Pennsylvania. It contains evidence for an occupation significantly older than those associated with Clovis points and has been used as an argument against the "Clovis-first" model. However, the dating of this site and its connections to other sites remain controversial.

Meteorite: An object that has fallen from outer space and is now on earth.

Microwaves: A form of electromagnetic radiation with wavelengths longer than 1 millimeter and shorter than 10 centimeters.

Middle Kingdom: Period of ancient Egyptian history when some of the classic works of ancient Egyptian literature were written. This time period is dated to about 2000–1800 BCE using in part a record of the heliacal rising of Sirius.

Monte Verde: An archaeological site in southern Chile considered one of the best candidates for a pre-Clovis occupation, although the age of this site has been disputed. The site also contains the remains of plant material, including medicinal plants and possibly occupation structures, that provide important information about how people lived in the New World at the end of the Ice Age.

Monotreme: A mammal that lays eggs.

Mutation: A change in the sequence of nucleotides in a DNA molecule; includes insertions, deletions, duplications, and substitutions.

New Kingdom: Period of ancient Egyptian history that includes the reigns of the famous kings Tutankhamen and Ramses the Great. This period is dated to about 1600–1100 BCE using, in part, a record of the heliacal rising of Sirius.

Neutrino: A neutral, very low-mass subatomic particle that participates in certain nuclear reactions such as beta decay.

Neutron: A massive, neutral subatomic particle found in the nuclei of atoms. Atoms with the same numbers of protons but different numbers of neutrons have the same chemical properties but different masses (see Isotopes).

Nucleus: The dense inner core of atom, where the protons and neutrons are located.

Nuclear Decay: The transformation of a nucleus from one element or isotope into another. Typical forms are known as alpha, beta, and gamma decay.

Nucleotide: A set of small molecules, including adenine, thymine, guanine, and cytosine, which form base pairs in DNA molecules.

Nobel Gases: Elements that only rarely form chemical bonds with other elements, including helium, neon, and argon.

Nuun Ujol Chaak: The ruler of Tikal who battled with, and was eventually defeated by, Yuknoom Ch'een and his allies.

Old Kingdom: Period of Ancient Egyptian history during which the Great Pyramids were built. This period covers some 500 years around 2500 BCE, but the exact dates are still uncertain.

Order: A piece of biological nomenclature that is used to identify a group of animals that share common heritage. Living mammals fall into about twenty distinct orders.

Ordinary Matter: Matter that is made up of atoms, nuclei, and electrons (as opposed to dark matter).

Orrorin tugenensis: A fossil hominid found in Kenya. Dating back to about 6 million years ago, this animal may hold important clues about when our ancestors first started to walk on two legs.

Parallax: Method of measuring distance to nearby stars that uses the apparent motion of the stars over the course of a year.

Phase Transition: A phenomenon that occurs when a material changes between different phases such as solid, liquid, gas, or plasma.

Placentalia: A group of mammals consisting of the last common ancestor of all modern placental mammals, and all of that animal's descendants.

Placental Mammal: A mammal in which the fetus is nourished by a placenta inside its mother.

Plasma: A state of matter in which atoms are ionized, so electrons and nuclei can move independently of one another.

Point Substitution Mutation: A type of mutation where a single base pair is replaced by another base pair, such as when the sequence ATGTG becomes ATCTG.

Potassium-40: An unstable isotope of potassium with 19 protons and 21 neutrons, which can decay into either calcium-40 or argon-40. The half-life is 1.28 billion years. Used in potassium-argon dating.

Potassium-Argon Dating: A method of dating volcanic deposits by measuring their potassium and argon contents. Assuming all of the argon-40 is due to the decay of potassium-40, we can determine how much potassium-40 has decayed since the rock solidified. Then, using the half-life of potassium-40 (1.28 billion years), we can estimate how long ago the lava cooled.

Power Spectrum: A graph that shows how much a signal on the sky varies as a function of the apparent size or angular scale of the variation.

Photon: A particle of light.

Phylogenetic Tree (Dendrogram): A graph that illustrates the relationships among different organisms.

Precession: The circular motion of the orientation of a spinning object's axis when the object is subjected to asymmetric forces. The earth undergoes precession due to interactions with the sun and the moon, which cause the orientation of earth's pole to move around in a circle once every 26,000 years.

Prior: An assumption used in Bayesian statistics that specifies the probability that a given theoretical prediction could have occurred prior to us making a measurement.

Proconsul: A fossil ape-like creature that lived twenty million years ago. It shares characteristics with all living great apes, but has no traits particular to any one of them. It therefore represents a branch of the ape family that diverged before any of the branches leading to the modern great apes.

Protein: A molecule composed of a string of amino acids. The sequence of amino acids determines the chemical properties of the molecule. Proteins participate in nearly all of the things a cell does, and the cell's DNA contains instructions for making many proteins.

Proton: A massive, positively charged subatomic particle found in the nuclei of atoms. The number of protons in the nucleus determines how many electrons surround it and thus establishes the chemical properties of the atom.

Quintessence: See Dark Energy.

Radiocarbon Dating: See Carbon-14 Dating.

Red Giant: A star that has exhausted the hydrogen fuel in its core and has consequently become larger, brighter, and redder than a typical main sequence star.

Redshifts: The shift to longer wavelengths in the light from distant galaxies. Possible mechanisms for producing redshifts include universal expansion and Doppler shifts.

Refractory: Melting or vaporizing only at high temperatures. The opposite of "volatile."

Rubidium-87: An unstable isotope of the element rubidium with a half-life of about fifty billion years, used to date ancient rocks like meteorites.

Sahelanthropus tchadensis: A fossil hominid found in Chad. Dating back to six million years ago, this creature could provide important clues about the origin of bipedalism in hominids.

Scale Factor: The distance between any two free (not gravitationally or electromagnetically bound) objects at any point in the past, divided by the distance between those same objects today. Used as a conventional measure of the "size" of the universe in efforts to track universal expansion.

Schonberg-Chandrasekhar Limit: A calculated limit to how much hydrogen an idealized star can burn before it will convert into a red giant. This limit depends on the mass ratio of hydrogen to helium and works out to be about 10%. Strictly speaking, this calculation does not apply to real stars.

Silent Mutations: Mutations that do not affect the structure or production of proteins.

Sivapithecus: A fossil animal that lived some twelve million years ago. It has features in its face, such as close-set eye sockets, that indicate a close relationship with living orangutans.

Solar Cycle: A roughly eleven-year period observed in the distribution and number of sunspots. The origin of this phenomenon is still not completely understood, but it appears to be correlated with global changes in the sun's magnetic field

Sothic Cycle: The 1,460-year cycle that resulted from the fact that Egyptians had a strict 365-day year with no leap days, which meant astronomical events occurred one day earlier each four years. It therefore took 1,460 years before an event like the heliacal rising of Sirius would again occur on the "proper" day. ("Sothis" is the Greek version of the Egyptian name for Sirius.)

Spectrum: The brightness of an object as a function of wavelength.

Stony-Iron Meteorites: Meteorites that contain a mix of metallic alloys and silicate minerals.

Stony Meteorites: Meteorites that are composed primarily of silicate minerals. Divided into two broad groups: chondrites and achondrites.

Sunspots: Dark blemishes on the surface of the sun. They appear to be associated with regions of intense and complicated magnetic fields. The number of sunspots rises and falls roughly every eleven years in a period known as the solar cycle.

Supernova: A dramatic explosion that marks the death of certain stars. A supernova may have initiated the formation of the solar system. One type of supernova also provides evidence for how the universe has expanded over the last ten billion years.

Tikal: The modern name of an important Classic Mayan city, originally known as *Mutul.* Calakmul's biggest rival.

Type Ia Supernova: One type of supernova that lacks hydrogen features in its spectrum. It is believed to occur when nuclear reactions restart in a white dwarf due to the accretion of additional material. These events provide a powerful tool for measuring the distances to far away galaxies.

Tzolk'in: A 260-day cycle that is part of the Mayan Calendar Round. A Tzolk'in date consists of a number between 1 and 13 and one of twenty day signs. With each passing day, the number increases by one and the day sign changes.

Vacuum Energy: An energy associated with empty space, one possible candidate for the dark energy.

Volatile: Vaporizing at relatively low temperatures. The opposite of "refractory."

Wavelength: A distance between adjacent crests in a wave. The wavelength of an electromagnetic light wave determines its color.

White Dwarf: A stellar remnant produced when the nuclear fuel is exhausted in a low-mass main sequence star.

Xenarthra: A primarily South American order of mammals that includes anteaters, armadillos, and sloths.

Younger Dryas: A period of time when the climate, after beginning to warm up after the end of the last Ice Age, suddenly cooled off again, at least in the Northern Hemisphere. It started around 12,700 years ago and lasted about 1,000 years. After this cold period, temperatures rose again and stabilized at present values.

Yuknoom Ch'een: An important ruler of the Mayan city of Calakmul during the Classic period.

Zalambalestids: A group of eutherian mammals that lived about seventy-five to ninety million years ago. Some features in the teeth of these creatures have led a few paleontologists to argue that they are related to modern rodents and rabbits, but this idea is highly controversial.

Zhelestids: A group of eutherian mammals that lived around eighty-five to ninety million years ago. Some characteristics in these creatures have been used to suggest a close link between them and modern hoofed animals. This idea is still controversial.

INDEX

Page numbers in italics refer to figures.